Recent Trends in INVERTEBRATES

Vol. 5

H.S.Bhamrah
Kavita Juneja

ANMOL PUBLICATIONS

ANMOL PUBLICATIONS
4378/4B, Ansari Road,
Darya Ganj, New Delhi - 110002.

Recent Trends in Invertebrates

ISBN 81 7041 473 3 (Set)
ISBN 81 7041 474 1 (Vol.I)
ISBN 81 7041 475 X (Vol.II)
ISBN 81 7041 476 8 (Vol. III)
ISBN 81 7041 477 6 (Vol. IV)
ISBN 81 7041 478 4 (Vol. V)
ISBN 81 7041 479 2 (Vol. VI)
ISBN 81 7041 480 6 (Vol. VII)
ISBN 81 7041 481 4 (V0l. VIII)

PRINTED IN INDIA

Published by J.L. Kumar for Anmol Publications and Laser Typeset by D.V. Computers, Delhi, Printed at Efficient Offset Printers, New Delhi.

PREFACE

The present title has been designed to meet the requirements of the undergraduate students. In the Recent Trends in Invertebrate all phyla have been discussed. The representative types of each phylum have been described in detail. This is followed by the classification of the phylum with a description of its common animals. Recent findings and work have been incorporated in the text. Special care has been taken to make the language as simple and lucid as possible. Simple and accurate diagrams accompany the text. In the interest of the students efforts have been made to cover the syllabi of nearly all Indian Universities.

The book features both a text and a laboratory guide. It deals with the structures and physiological phenomenon of each system. General topics are also dealt in details. It is hoped that this book will not only meet the requirement of undergraduate students but also be useful as a guideline to the teachers in their teaching. It is also hoped that the general public will find that the book is not riddled with too much of scientific and technical jargon.

It is difficult to acknowledge adequately the assistance of all who have contributed to the preparation of this book. First of all the authors express their thanks to their friends and colleagues whose continuous inspiration have initiated them to bring out the present volume.

The authors are grateful to Dr. J. Bahadur, Professor and Head, School of Studies in Zoology, Jiwaji University Gwalior, Dr. A.K. Pandey, Professor and Head, School of Studied in Zoology, Vikaram University, Ujjain for their valuable suggestions.

Constructure criticism from students and teachers will be highly appreciated by the authors.

H.S. Bhamrah

Kavita Juneja

CONTENTS

1

CHARACTERS & CLASSIFICATION OF ANNELIDA

All the invertebrate animals, who possessed a soft-body, usually bearing a general resemblance to the common earthworm were included in the old phylum Vermes and were called worms. Thus the old phylum Vermes represented a heterogenous assemblage of no exact classificatory value. *Cuvier* was first who pointed out the fundamental differences between the lower unsegmented worms and the higher segmented worms in 1798. Now-a-days lower unsegmental worms divided into three phyla - Plathelmintches (includes flat worms), Nemathelminthes (round worms) and Trochelminthes (wheel animals). *Lamarck* gave the name Annelida (ring worms) to the higher segmented worms.

Term Annelida is derived from two words one is Latin (*annulus* = a ring) and a Greek (*eidos* = form).

Annelids are typical animals that burrow or crawl upon the bottom in the ocean or in fresh water. Some are free swimming, at least during the breeding season, and others are adapted to life in moist soil. Many marine species build permanent tube which they inhabit ; in other species the tube is temporary, like a burrow, and may be abandoned for another. Some annelids are actively predaceous, seizing their prey in well developed jaws, others notably permanently tube dwelling polychaetes, feed upon microscopic particulate matter brought in by ciliated tentacles or entrapped in sheets of mucus.

Characters :

1. Annelids are bilaterally symmetrical with elongate vermiform body.
2. The body is metramerically segmented and ringed appearance whith the serially repeated segments give to the body. The body is composed of segments which are usually of similar structure internally as well as externally.
3. Presence of cuticles secreted by the underlying epidermis containing unicellular gland cells and sensory cells.
4. Appendages unjointed and are minute, rod-like, chitonous setae, few to many per somite generally absent in *Hirudinea.*
5. Body wall consists of layers of circular and longitudinal muscles.
6. Coelom well developed, extensive, separated into segmental compartments by transverse partitions, or septa, but coelom is reduced in Hirudinaria.
7. The digestive tract complete, tube like, extending entire length of body from mouth to anus.
8. A closed circulatory system of longitudinal blood vessels with lateral branches in each somite, blood plasma contains dissolved haemoglobin and free amoebocytes.
9. Respiration by epidermis, by gills in some tube dwellers.
10. Excertion by segmental paired nephridia, each removing wastes from the coelem and blood stream directly to the exterior.
11. Nervous system with ventral nerve cord composed of paired ganglia at segmental intervals connected anteriorly with the dorsal "brain" by a pair of circumpharyngeal connectives.
12. Sensory organs in the form of sensory cells, organs for touch, taste and light preception are prseent.
13. In few the sexes are separate, while in others sexes are united (hermaphrodite).
14. Gonads may be simple structure as in *Nereis*, where they develop only during breeding season. In some forms gonads are well developed.

15. Cleavage is spiral and determinate.
16. Development direct or indirect through a larval stage, the *trochophore.*
17. Process of regeneration is common.

Classification

There are about 8,700 known species of segmented worms which are divided into four main classes. The classification is primarily based on the basis of presence or absence of setae, parapodia and metamerism. The 4 classes are :

Class - I. Polychaeta

Class - II. Oligochaeta

Class - III. Hirudina

Class - IV. Archiannelida.

CLASS. I. POLYCHAETA

(Gr. *poly* = many ; *chaete* = bristles).

1. These are marine and carnivorous annelids.
2. Body usually elongated, cylindrical and distinctly segmented into numerious similar metameres.
3. Head consists of prostomium and peristormium and bears eyes, tentacles, cirri and palps.
4. Setae are numerous and are borne up on lcteral prominences of the body wall known as parapodia.
5. Clitellum is absent.
6. Parapodia highly vascular and respiratory in function.
7. Coelom is specious usually divided by intersegmental septa.
8. The blood vascular system is well-developed and does not communicate with coelom.
9. Alimentary canal is provided with an eversible buccal region and protrusible pharynx.

10. Sexes are separate. Gonads are simple and developed by active proliferation of coelomic epithelium.
11. Fertilization external, free-swimming trochophore larva.
12. Asexual reproduction by lateral budding.

Class Polychaeta is divided in two orders :

Order (i) Errentia

1. Free swimming, often pelagic while some are tubicolous (live in tubes).
2. All segments are similar except head and regions.
3. Parapodia provided with cirri, are equally developed throughout. These are organs of locomotion.
4. The head is usually definite which bears eyes, tentacles and cirri.
5. Pharynx is usually protrusible and armed with chitinous jaws and teeth.
6. Clitellum absent.
7. The branchiae or gills are present but not restricted to anterior end of body.

Examples: *Aphrodite, Polynoe, Nereis, Glycera, Eunice, Syllis*

Order (ii) Sedentaria

1. Sedentary in habit. Burrowing and tube dwelling.
2. Body is distinguished into two or three regions due to differences in shape of segments, parapodia and setae.
3. Head is small or much modified, without eyes and tentacles, prostomium indistinct.
4. Gills are situated only in the anterior part of the body, when present.
5. Pharynx is non-protrusible devoid of jaws or teeth. Feeding on plankton or organic detritus.
6. The parapodia are peculiar which are confined in the posterior region of body and without cirri.

Examples : *Chaetopterus, Sabella, Arenicola, Terebella, Amphitrite, Serpula, Spirobis.*

CLASS. II. OLIGOCHAETA

(L., *oligos* = few ; *chaete* = bristles)

1. Mostly terrestrial, some live fresh-water.
2. Body with conspicuous external and internal segmentation.
3. Head not distinct, prostomium is usually small, without eyes and tentacles.
4. Parapodia and cirri are absent. Setae are usually arranged segmentally.
5. Clitellum usually present.
6. Pharynx is not protrusible and without jaws.
7. Hermaphrodite i.e. bisexual. Gonads are complicated. Overies and testes are compact structures.
8. Development direct without larval stage.

Class Oligochaeta is divided in two orders :

Order (i) Archioligochaeta

1. Mostly fresh water forms.
2. Body has few segments. Setae present.
3. Gizzard is either absent or poorly developed.
4. Simple clitellum, consists of single layer of cells situated far forwards.
5. Eye spots are usually present.
6. Reproduction is both asexual and sexual.
7. Male reproductive openings lie in front of female reproductive openings.

Examples : *Tubifex, Actosoma.*

Order (ii) Neooligochaeta

1. Mostly terrestrial forms found in moist soil.
2. Body is large with many segments.
3. Setae are arranged in lumbricine manner.
4. Well developed gizzard.
5. Clitellum after 12th segment and is composed of two or more layers of cells.
6. Female genital apertures before male genital pore and usuallly lie in 14th segment.
7. Eye spots are developed.
8. Reproduction sexual only.
9. No free larval stage.

   Examples : *Pheretima, Lumbricus, Megascolex, Eutypheus.*

CLASS III. HIRUDINEA

1. They are marine, aquatic, terrestrial or free living. Some are ectoparasitic and blood-sucking annelids.
2. Body is segmented and is marked externally by ring-like annuli.
3. Presence of anterior and posterior cup-shaped suckers which are the organs of adhesion and locomotion.
4. The ventral mouth lies in the cup-shaped hollow cavity of anterior sucker. The anus is dorsal and situated just above the posterior sucker.
5. The coelom is greatly reduced due to the presence of connective tissue and is represented by haemocoelomic sinuses.
6. The true blood vessels with definite muscular walls present.
7. The nervous system includes a brain united by small oesophageal connectives to a double ganglionated ventral nerve cord.
8. Nephridia are segmentally arranged and open independently to exterior by nephridiopore, which are the organs of excretion.

9. Hermaphrodite, the testes are many and segmentally arranged, whereas ovary is single paired. Eggs are laid in cocoons.
10. Fertilization internal, development direc wiothout larval stage.

Class Hirudinea is divided into following orders :

Order (i) Acanthobdellida

1. Mostly ectoparasite on the fin of salmon fishes.
2. Anterior sucker wanting, posterior sucker well formed.
3. Proboscis is small and not well developed.
4. Nephridia without internal openings.
5. Body generally composed of twenty or twenty-one segments.
6. It forms a connecting link between Oligochaeta and Hirudinea.

Example : *Acanthobdella*.

Order (ii) Rhynchobdella

1. Aquatic leeches, parasite on snails, frogs and fishes.
2. The anterior part of body may be protruded and retracted so as to form a proboscis or introvert. Jaws and absent.
3. Each typical body segment consist of 3, 6 or 12 rings.
4. The mouth is minute slit-like opening placed inside the anterior sucker.
5. Blood is colourless. Coelom is reduced to sinuses without botryoidal tissues.

Examples : *Branchellion, Glossiphonia, Pontobdella, Piscicola.*

Order (iii) Gnathobdella

1. Fresh water and terrestrial forms. Ectoparasites and blood sucking.
2. Each typical segment consists of five rings or annuli.
3. Anterior sucker has three jaws, one median dorsal and two ventro-lateral.

4. Proboscis non-protrusible or absent.
5. Blood is red in colour.
6. Coclom is reduced. Botryoidal tissue present.

   Examples : *Hirudo, Hirudinaria, Haemopis, Aelostoma.*

Order (iv) Pharyngobdella

1. Freshwater amphibious carnivorous leeches.
2. Each body segment is divided into five rings or annuli.
3. Pharynx non-protrusible. Jaws are absent.

   Examples : *Dina, Erpobdella, Trocheta.*

CLASS. IV ARCHIANNELIDA

1. This group represents the primitive members of Annelida.
2. Mostly marine worms.
3. Body is elongated, cylindrical with internal segmentation.
4. The head is distinct bearing the tentacles, eyes and ciliated pit.
5. Parapodia are wanting.
6. Hermaphrodite or unisexual; gonads are simple and develop temporarily only during breeding season by active proliferation of colomic epithelium.

About 45 species, one dozen genera are included in this class. The class is divided into five families.

(i) Polygordidae *e.g. Polygordius. Protannelis.*

(ii) Protodrilidae *e.g. Protodrilus.*

(iii) Saccocirridae *e.g. Saccocirus.*

(iv) Dinophilidae *e.g. Dinophilus, Diurodrilus, Trilcbodrilus.*

(v) Nerillidae *e.g. Nerilla, Nerillidium, Troglochaetus.*

POLYNOE

Phylum	-	Annelida
Class	-	Polychaeta
Order	-	Errantia
Genues	-	*Polynoe.*

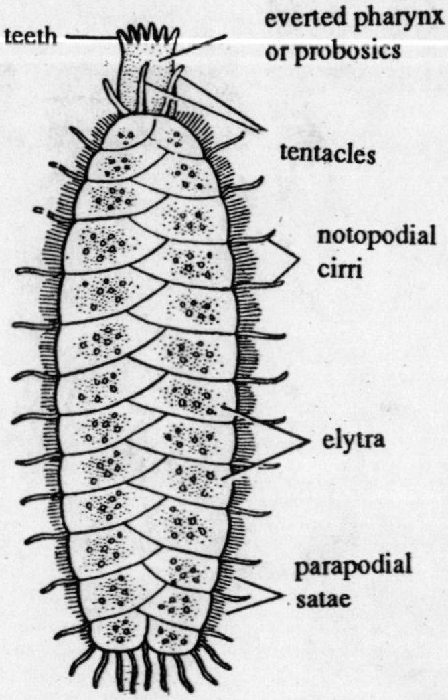

Fig. 1.1 *Polynoe*

1. It is a bioluminescent polychaet found under stones in sea-water near tide mask.
2. It is found in U.S.A. (South Carolina to Massachusetts) and Europe.
3. Body is short, broad and oval with few segments. Body shows considerable degree of dorso-ventral compression.
4. It has a covering of elytra which are modified dorsal cirri. Elytra do not contain setae and have rich nerve supply.
5. Elytra are bioluminescent and the illuminating elytra are cast off when distrubed or elytra are shed-off to decive the enemy.
6. Head bears three tentacles and two pairs of peristomial cirri.

7. Pharynx is protrusible.
8. It is carnivorous and may even eat each other (cannibalism).
9. Fertilizad eggs and resulting embryos adhere in masses to the dorsal surface under the shelter of elytra.

EUNICE

The classification is same as that of *Polynoe*.

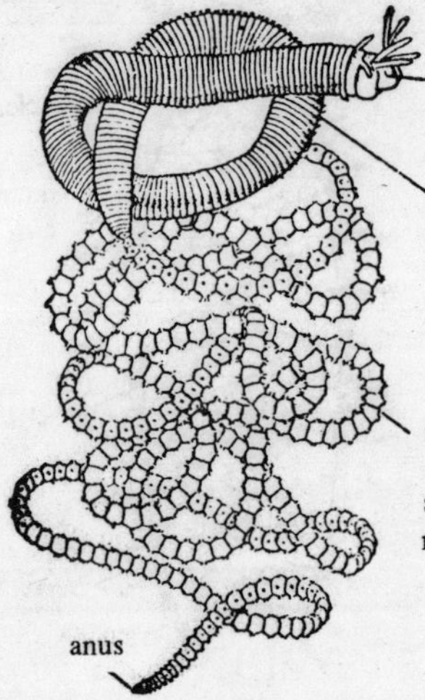

Fig. 1.2 The Pacific palolo worm, *Eunice viridis*

1. It is tube-dwelling marine polychaete found among the coral reefs at sea-bottom.
2. It is found in Atlantic and Pacific oceans, West Indies and Fiji Islands.
3. It is commonly known as 'palolo-worm'.
4. Body elongated with numerious segments. Prostomium consists of two segments with one pair of cirri, five tentacles and a large palp.
5. Gills begin usually on 6th segment.
6. Body yellowish or brownish and differentiated into anterior

sexless *atoke* and posterior sexual *epitoke* regions. Epitoke projects from the tube and gets separated from atoke, and come to the surface, swim vigorously and finally discharge the gonads. The empty epitoke sinks to the bottom after swarming.

7. Fertilized eggs develop into a larval form.

SYLLIA

The classification is the same as that of *Polynoe*.

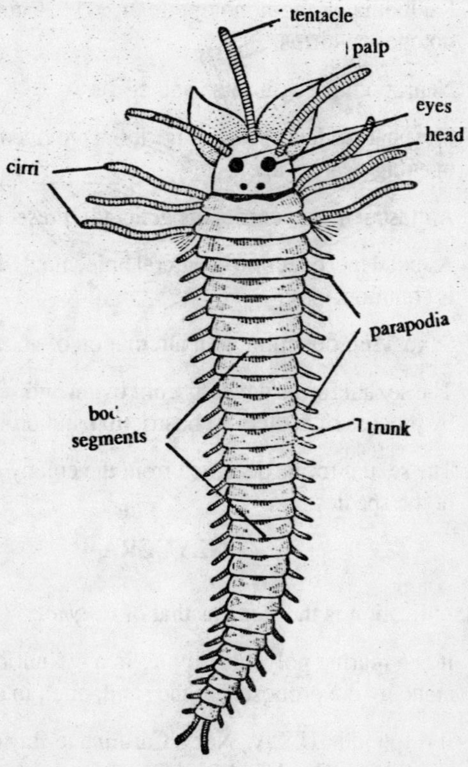

Fig. 1.3. Syllis

1. *Syllis* is a pelagic, marine polycheat.
2. *Syllis* is common in U.S.A., Woods Hole regions, Long Island, Sound to Bay of Fundy.

3. Small elongated worms, less than an inch in length.
4. Head small with prostomium and peristomium.
5. Prostomium with eyes, three tentacles and two fused palps.
6. Peristomium carries two pairs of cirri, tentacles and cirri are fused.
7. Each segment of the body carries a pair of parapodia.
8. Parapodia without notopodium, represented by long jointed notopodial cirrus.
9. Neuropodium with setae and cirrus.
10. The pharynx has a single large tooth provided with a poison gland opening near its apex.
11. A muscular proventriculus generally present.
12. Asexual reproduction by lateral branching, fission and gemmtion is common.
13. Sexual reproduction with alternation of generations.
14. The sexual forms develop from ovum and sexual forms develop by process of posterior constriction and proliferation.
15. The sexual forms detached from the colony and help in dipersal of the species.

GLYCERA

The classification is the same as that of *Polynoe*.

1. It is a marine polychaet living in a cylindrical burrow which is made by the proboscis in the sand, mud, in shallow water.
2. It is found in U.S.A., North Carolina to Bay of Fundy and South Carolina to Cape Cod.
3. The boby of *Glycera* is elongated, cylindrical and meta merically segmented.
4. The head is distinct bearing a small conical prostomium with minute tentacles and rudiment palps.
5. The parapodia are small out-growths of integument which contain special gills.

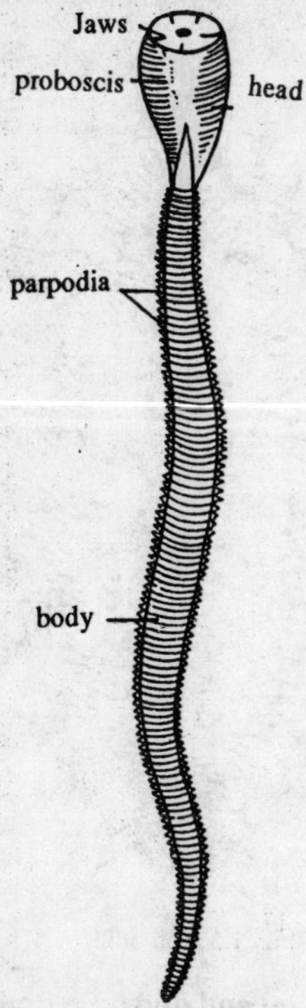

Fig. 1.4 Glycera.

6. Presence of a long, tubular proboscis bearing teeth, which give the peculiar shape to the worm.
7. There are no blood vessels; each gill contains a diverticulum of coelom. However, the perivisceral fluid contains numerous red corpuscles which assume the function of blood.
8. It is generally used as a fishbait.

APHRODITE

1. *Aphrodite* or sea mouse commonly found in deep water 10 to 100 fathoms in muddy bottoms of almost all seas.
2. It generally burrows in mud and creeps on the sea bottom.
3. Fairly common in coral line.
4. It is commonly known as ''sea mouse''.

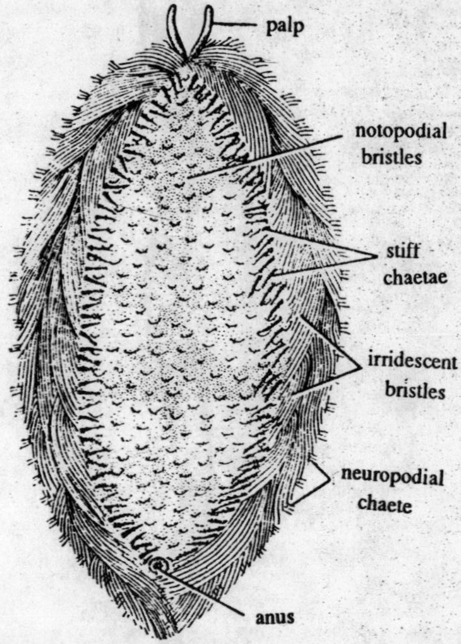

Fig. 1.5 Aphrodite

5. Body short almost oval and dorso-ventrally compressed.
6. Head small generally hidden, situated at the anterior end.
7. The head carries a single median tentacle and two large lateral palps.
8. Body is covered by modified dorsal cirri the elytra, which are plate like.
9. The elytra is a respiratory organ.
10. Many of the notopodial chaetae are iridescent forming a blanket of felt like covering of the whole body of animal.
11. Apart from these brilliantly coloured thread-like setae three are spine-like brown stiff bristles.
12. The anus is situated at the extreme postrior end on the dorsal surface.
13. When distributed the stiff irdiscent bristles can be erected just

like the spine of porcupine to frighten away the enemies.

14. The alimentary canal has caeca for the digestion of food.
15. When animal moves in water it shows beautiful iridescent colour.

TROMOPTERIS

The classification is the same as that of *Polynoe*.

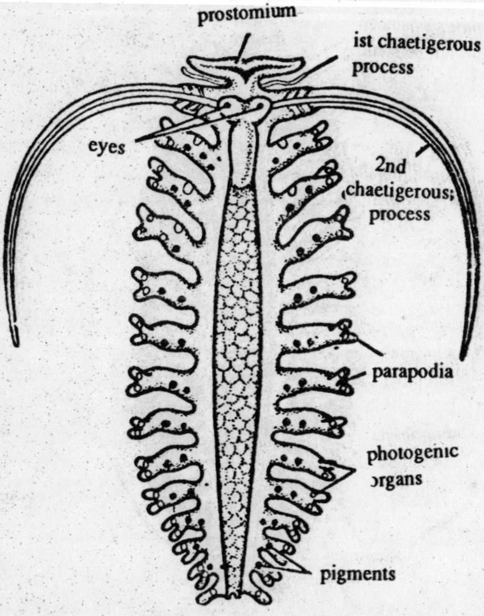

Fig. 1.6. *Tomopteris*.

1. It is a marine pelagic polychaete.
2. It is found in U.S.A. and Europe.
3. Body is colourless, transparent, having 18-20 segments.
4. Parapodia are large, bilobed and without setae.
5. Each lobe of parapodia bears a yellowish rosette-shaped or spherical photogenic organ.
6. The head is distinct due to the presence of hammer-shaped

prosomium which contains a pair of eyes and two pairs of chaetigerous processess or tentacular cirri. The anterior are very short which posterior ones are very much elongated.

7. The ventral mouth is without jaws.
8. Circulatory system is absent.

MYZOSTOMA

The classification is the same as that of *Polynoe.*

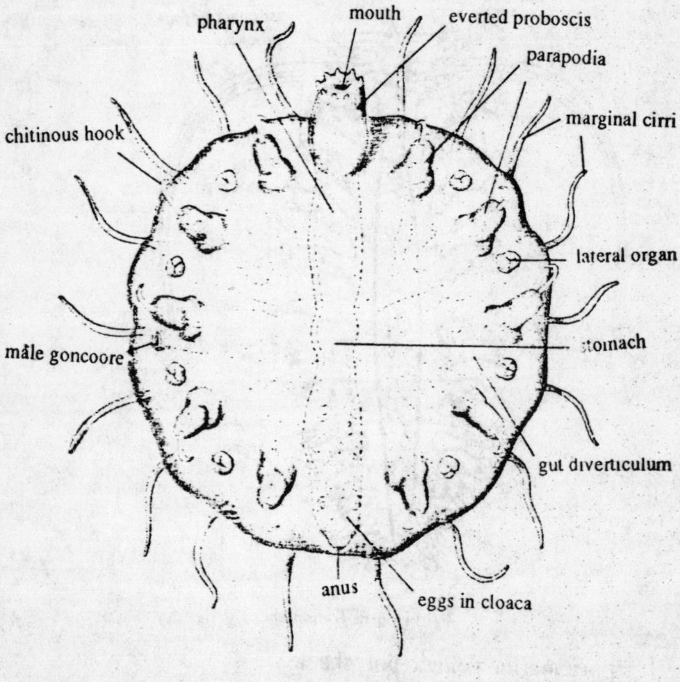

Fig. 1.7

1. It is an unsegmented, flatenned polychaete, generally ectoparasites on the starfishes, sea-lilies or feather stars.
2. They are found in West Indies, Havana, Europe.
3. These are disc-shaped situated on convex ventral side.
4. Five pairs of parapodia situated on convex ventral side.

5. It has four pairs of small suckers, probably sense organs, on ventral side.
6. Ten pairs of minute protuberances of short cirri extend from the edge of the body.
7. Mouth is ventral in position that leads into a protrusible pharynx.
8. A distinct head is absent.
9. Coelom or body cavity is absent.
10. Circulatory system and respiratory organs are absent.
11. Hermaphrodite,protandrous. Later testes degenerate leaving behind female reproductive organs.
12. Development indirect through trochophore larva.

CHAETOPTERUS

Phylum	-	Annelida
Class	-	Polychaeta
Order	-	Sedentaria
Genus	-	*Chaetopterus.*

1. It is a tubicolous worm which generally lives permanently in a 'U' tube, made up of sand and mucous with excurrent and incurrent openings.
2. It is found in Europe, U.S.A., North Carolina to Cape Cod.
3. It is commonly as 'paddle-worm'.
4. Long worms, about 0.15 to 0.45 metre in length.
5. Body divided into three regions :

(*a*) Anterior flattened region with a funnel-shaped peristomial collar having two tentacles called peristomial cirri. In the 4th segment setae are enlarged, while 10th segment has a pair of wing-like aliform notopodia.

(*b*) Middle region possesses modified parapodia forming three pairs of fans. It consists of 5 segments.

(*c*) Hinder or posterior region has smaller segments bearing parapodia. These segments are about 30 in number.

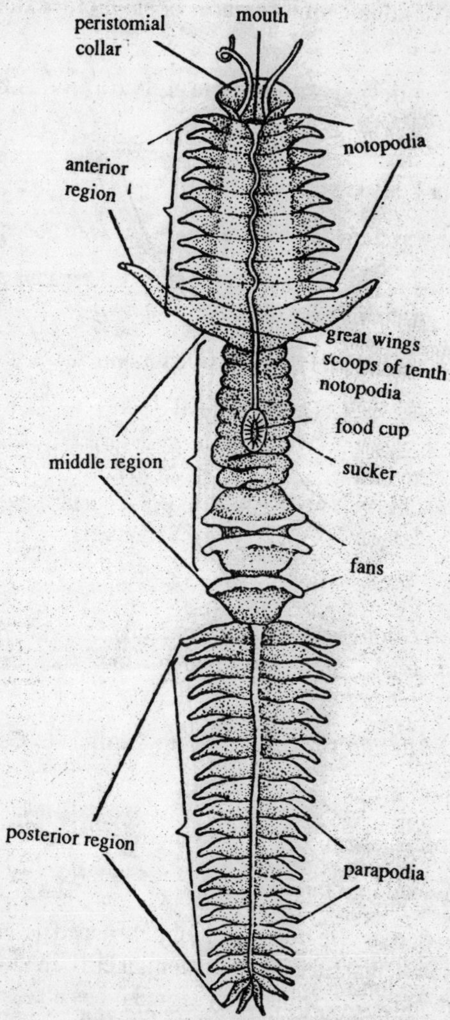

Fig. 1.8 *Chaetopterus*. Entire worm.

6. Feeding is very characteristic. Three paired fans of middle region and cilia of ciliated groove create a water-current, which flows from one end of the tube to the other. The water-current serves for nutrition and respiration. Notopodia secrete mucus, which forms mucous begs with food particles. It is rolled into a ball or bolus which is passed on to the mouth through dorsal groove.

7. *Chaetopterus* is highly bioluminescent emits blue-green light.
8. Reproduction is usually asexual by transverse fission.
9. A great power of regeneration, the whole body can be regenerated from a single segment.

ARENICOLA

The classification is the same as that of *Chaetoplerus*.

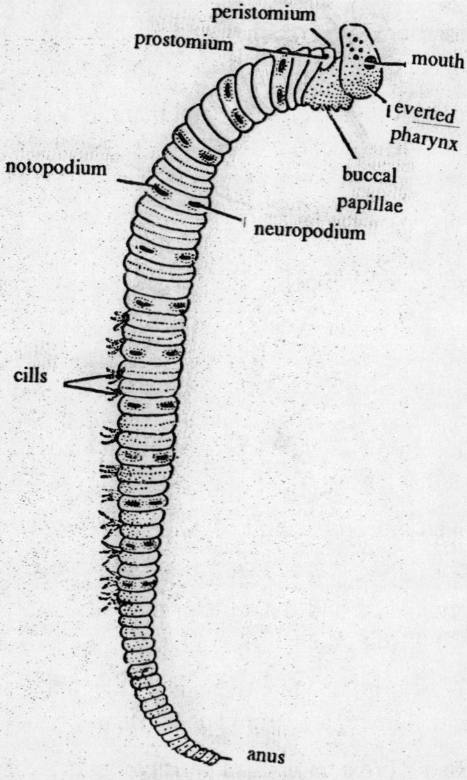

Fig. 1.9 Arenicola.

1. It is found on coasts between tide marks. It lives in a 'U' shaped burrow made up of sand and mucous.
2. It is found in Europe and U.S.A. (from Florida to Cape Cod).
1. It is popularly known as 'lug worm'.

2. The body is elongated, worm-like, broadest at anterior end.
3. The length of the worm is 5 to 8 inches.
4. The body is divided into three regions.
5. The anterior or prebranchial region of the peristomium, an achaetous segment and six chaetiferous segments. Eyes, tentacles absent. Peristomium contains 9 pairs of lithocytes.
6. The middle region not so thick of 12 or 13 chaetiferous and branchiferous segments.
7. The posterior region much thinner without setae and branchiae, with segments less clearly marked and variable in number.
8. Mouth lies ventral to the prostomium. The buccal region and pharynx protrusible as proboscis which is covered by chitinized papillae.
9. Nephridia six pairs.
10. It is generally used as baid for fishes.

AMPHITRITE

The classification is the same as that of *Chaetopterus*.

1. It is a tubicolous marine worm, living in tubes or burrows lined with sand and mucous.
2. It is found in Europe and U.S.A. (from North Carolina to Cape Cod).
3. Body elongated and cylindrical, pinkish in colour measuring about 20-30 cm in length.
4. Body is differntiated into three district regions : anterior region (head), middle region (trunk) and posterior region (abdomen).
5. Head small; prostomium and peristomimum are fused to form a horse-shoe shaped preoral lobe.
6. The prostomium forms the upper lip of and peristomium lower lip of the mouth. Eyes and palps are absent.
7. A tuft of long, yellowish and contractile tentacular filaments arises from a transverse ridge just behind the pre-oral lobe.

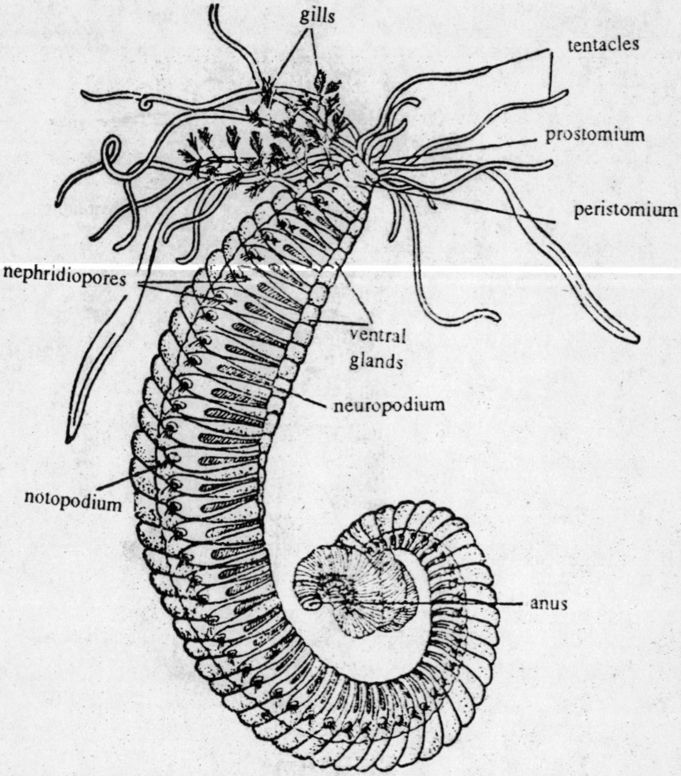

Fig. 1.10 Amphitrite.

8. Tentacles are feeding organs bearing ciliated groove on its oral surface.

9. The first three segments of thorax bear red and feather-like branched gills. Rest of the segments bear notopodia with setae. Some of the thoracic segments have shield glands for mucous secretion.

10. Abdomen is long and many segmented. Each segment contains notopodia and neuropodia.

11. The caudal segments lack notopodia and anus is found terminally in the last segment.

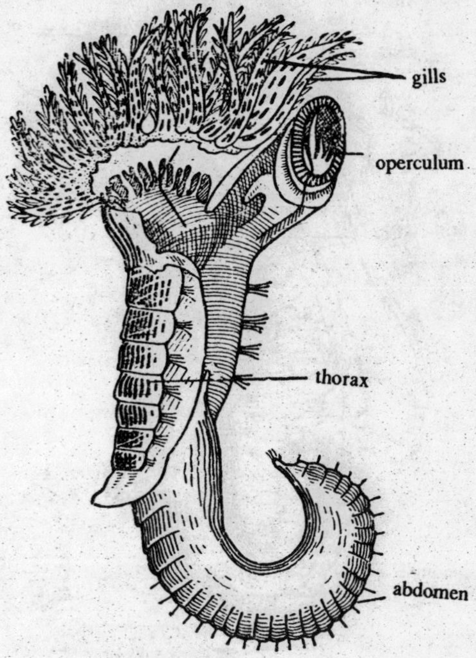

Fig. 1.11 Serpula.

SERPULA

The classification is the same as that of *Chaetopterus*.

1. *Serpula* is a tubiculous worm, generally lies in tubes formed of mucin hardened with calcarous matter.
2. The tubes are generally 1.50 inches long.
3. Found in Europe and U.S.A. and commonly called 'fan-worms'.
4. The worm is minute with small trunk and abdominal segments.
5. The prostomium carries a pair of large semi-circular feathered gills which are modified palps.
6. The gills are respiratory and also serve as food collecting organs.
7. One of the tips of gill filament is modified into stopper or

operculum which lock the aperture of tube when *Serpula* withdraws inside the tube.

8. The dorsal and ventral cirri of thoracic segments are fused to form a thoracic membrane on each sibe which is used to smoothen the interior of the tube.

9. The nephridia of posterior segments act as genital ducts.

SABELLA

The classification is the same as that of *Chaetopterus*.

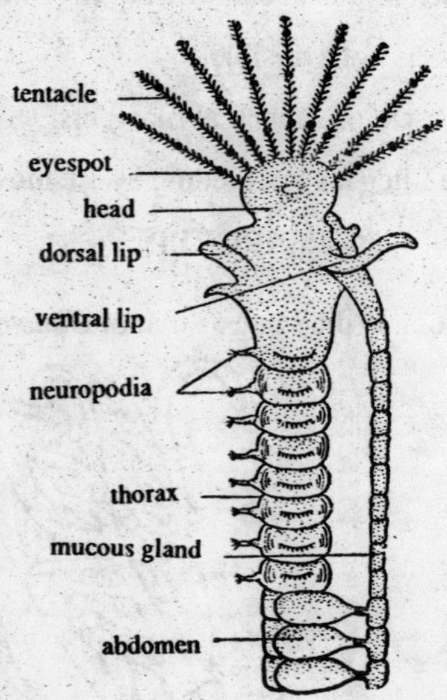

Fig. 1.12 Sabella.

1. Tubicolous polychaetae, live in tubes of sea-mud or sand which they rarely or never leave.

2. It is distributed throughout the world but especially U.S.A. (North carolina to Cape Cod).

3. It is commonly called as 'peacock-worm'.

4. Body is elongated bearing equal number of segments.

5. Body is divisible into a small head, thorax and abdomen.
6. Head consists of prostomium with several large gill-filaments or feathery tentacles which are modified palps. Each tentacle has two rows of small pinnules having mucous secreting cells.
7. Thorax formed of 8 segments and provided with gland-shields and parapodia. The gland secrete mucous. The first segment is called peristomium which is collor-like and has notopodium and setae but no neuropedium.
8. Abdomen is formed of 300 segments, and is marked with a faecal groove which serves to carry faeces out of the tube.
9. Buccal mass is not eversible.
10. Eyes are present on branchial filaments.
11. *Sabella* is a filter-feeder, feeding by the movements of tenacles.

SPIRORBIS

The classification is the same as that of *Chaetopterus*.

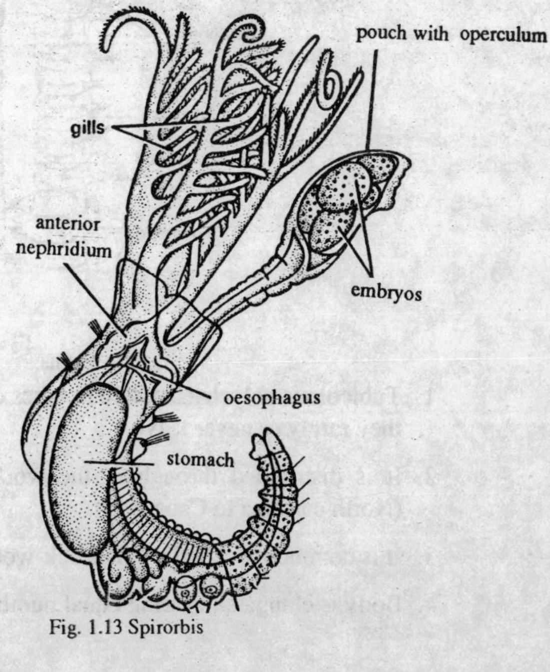

Fig. 1.13 Spirorbis

1. It is a minute tubicolous, sedentary polychaete, living in a small calcareous tube, coiled in a flat spiral manner.
2. It is commonly found in U.S.A., Long Island to Bay of Fundy and northwards.
3. Body measure 3mm in length.
4. The body is cylindrical, elongated and is divisible into upper thorax and lower abdomen.
5. Each gill consists of an elongated stem on which are borne the short filaments ; one of the gill filaments is enlarged to form an operculum which closes the mouth of tube.
6. The animal is hermaphrodite : anterior part of body gives rise to the ova while posterior part or abdomen produces the sperms.
7. Eggs are developed inside the operculum.
8. The prostomium and peristomium are compactly fused together to form a collar which secretes the tube.
9. Presense of paired anterior nephridia which are excretory in function.

STERNEASPIS

The classification is the same as that of *Chaetopterus.*

1. It is a marine, tubicolous polychaete.
2. It is found in U.S.A. (Florida).
3. The body is short, broad and made up of about 30 segments.
4. Head is formed of roistomium and peristomium.
5. Prostomium is elongated and without any appendages.
6. Peristomium is knob-like and contains mouth.
7. First three segments possess incomplete bristles, segments 8th-15th possess setae while segments 5th, 6th, 7th and 16th - 24th are without setae.
8. Intestine is coiled.

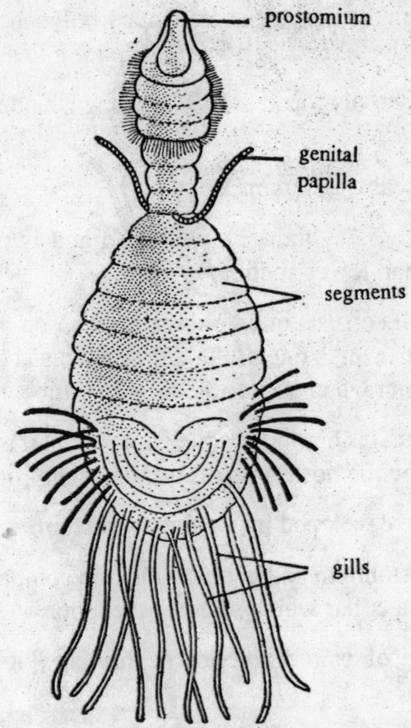

Fig. 1.14 Sternaspis

9. Anus is terminal and surrounded [illegible] gills.

10. Presence of a single pair of nephridia bearing small ciliated funnels, which do not open to exterior.

11. A ventral shield formed by a thickening of the cuticle in the posterior region of the body bears a number of setae around its edges.

TEREBELLA

The classification is the same as that of *Chaetopterus*.

1. It is a marine, burrowing, sedentary polychaete.

2. It is found in Europe, U.S.A., Woods-Hole region and Vineyard Sound to Bay of Fundy.

3. Body is long and vermiform differentiated into head, thorax and abdomen.

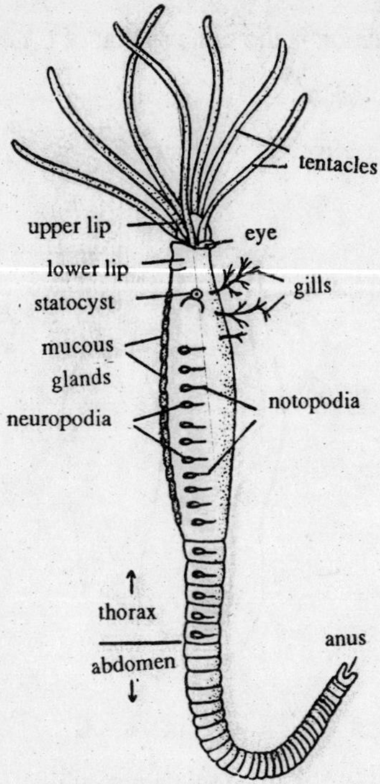

Fig. 1.15 Terebella.

4. Head is horse-shoe shaped made up of prostomium and peristomium.
5. Prostomium flatenned into a mobile upper lip which carries a transverse series of numerous filiform tentacles and two pairs of eyespots. Palps are absent.
6. Peristomium forms lower lip and lacks cirri.
7. The gills or branchiae are found on the dorsal surface of few anterior segments and they form four wide comb-like branches on a single peduncle.
8. Thorax with gland shields.
9. Parapodia are feebly developed.
10. Nephridia are large and small and serve as both excretory and genital ducts.

MYXICOLA

The classification is the same as that of *Chaetopterus*.

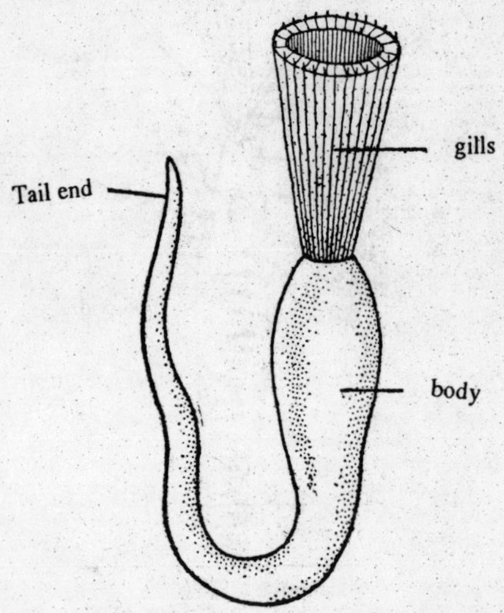

Fig. 1.16 *Myxicola*

1. *Myxicola* is a sedentary tubicolous animal lives in a tube in the sea sand. Occasionally the worm leaves the tube and swims freely.
2. Body is elongated, dark green in colour and measures about 10 cm in length.
3. Eyes, mucous and shields are absent.
4. Gills are present which are well developed and protrudes out.

AEOLOSOMA

Phylum	-	Annelida
Class	-	Oligochaeta
Order	-	Archioligochaeta
Genus	-	*Aeolosoma*

1. It is a freshwater worm found in ponds, pools and living among the algae.

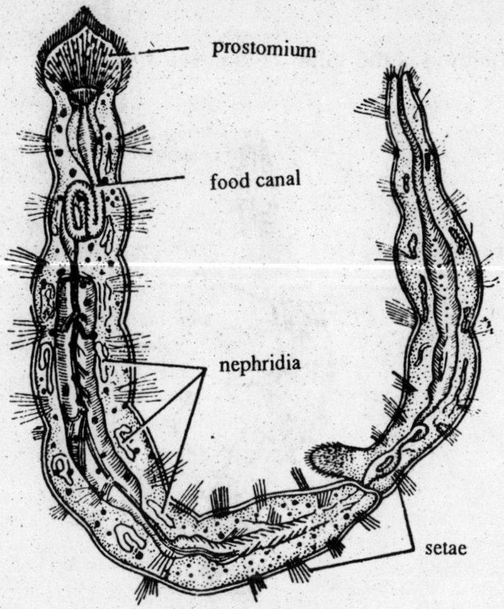

Fig. 1.17 Aelosoma.

2. Body transparent and dorso-ventrally flatenned and spotted with red, green or yellow oil globules in the integument.

3. Body is made up of 7-10 segments, which are not clearly marked. Coelomic septa are absent.

4. Prostomium ciliated ventrally and bears a pair of ciliated pit.

5. Each segment bears four bundles of hair-like setae.

6. Clitellum present only on the ventral side of the 5th - 7th segments.

7. Nephridia are paired and metamerically repeated.

8. Nervous system is hypodermic, the ventral nerve cord is absent.

9. These are hermaphrodite worms but commonly reproducing asexually by transverse fission.

10. Nephridial ducts act as gonoduct. The genital products pass out through nephridiopores.

TUBIFEX

The classification is the same as that *Aelosoma.*

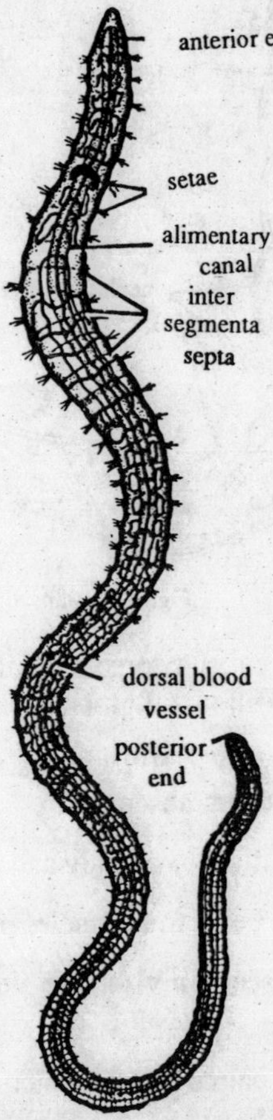

Fig. 1.18 Tubifex.

1. It is a tubicolous fresh water animal found on the bottom of deep lakes. It lives into tubes, made up of mud and minerals and mucous.
2. It is found in U.S.A. (Long Island Sound to Maine).

3. They occur in large numbers in the form of patches.
4. The body is elongated, somewhate cylindrical reaching upto the length of 4 cm. approximately.
5. It is reddish in colour and is divisible into a number of ring-like metameres.
6. Each segment of body except the few posterior segments bears the tufts of minute setae on its either side.
7. The posterior end of body protrudes out from the tube and waves back and forth.
8. The clitellum is found in 11th and 12th segments.
9. Contractile heart is present in 8th segment.
10. Male genital pore lies in the 11th segment and female genital pore in the 12th segment.
11. The eggs are deposited in cocoon, which contain food yolk for the nourishment of embryo. The epibolic type of process occurs during development.

LUMBRICUS

Phylum	-	Annelida
Class	-	Oligochaeta
Order	-	Neooligochaeta
Genus	-	*Lumbricus*

1. It is found in moist soil (terrestrial).
2. It is a common earthworm of Europe and U.S.A.
3. Body is cylindrical, 15-30 cm in length and made up of 100-150 segments.
4. The lobular prostomium divides peristomium into two parts.
5. Clitellum present in segments from 32-37 and is swollen.
6. Each metamer except first and last, bears 'S' shaped chitinous chaetae arranged in 4 pairs of which two are ventral and two ventro-lateral.

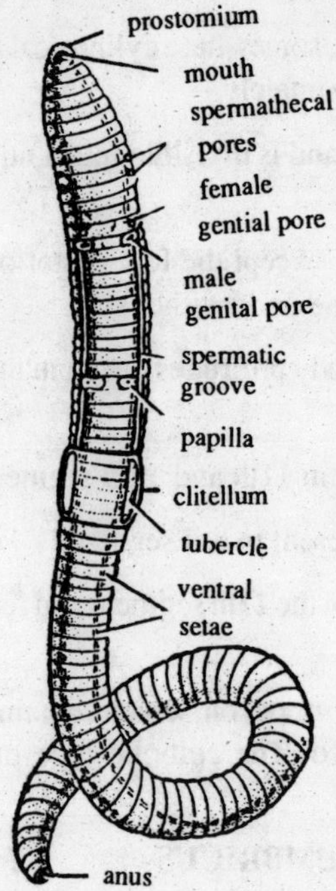

Fig. 1.19 Lumbricus.

7. Female and male genital pores are paired and are situated in 14th and 15th segments.
8. Gizzard is absent.
9. Prostate glands are absent.
10. Genital paillae absent.
11. These are two pairs of spermathecae in 9th and 10th segments. Three pairs of seminal vesicles in 9th, 11th and 12th segments. Testis sacs are two, one in 10th and other in 11th segment.
12. Each segment has a pair of nephridiopores, situated ventro-laterally.

MEGASCOLEX

The classification is the same as that *Lumbricus*.

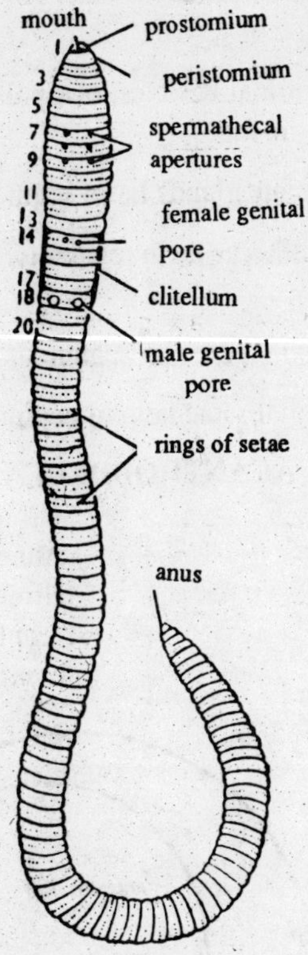

Fig. 1.20 *Megascolex.*

1. It is a terrestrial worm lives in moist soil.
2. It is found in South India, Sri Lanka, Australia, Tramania and New Zealand.
3. It measures 8-12 cm in length and bears 80-100 segments.
4. The dorsal body surface is dark purphlish brown and the ventral surface is somewhat pale in colour.
5. Clitellum present in segments from 14-17.
6. Gizzard lies in 6th segment.
7. Longitudinal blood vessels are two, dorsal and ventral and they are connected by eight pairs of hearts in segments 6-13.

8. One pair of metanephridia is present in each segment behind the 19th segment.
9. Two pairs of seminal vesicles, one pair lies in 9th segment and other in 12th segment.
10. One pair of prostate glands lie in segments 18 and 19.
11. 3 pairs of spermathecae lie in segments, 7, 8 and 9. Each with two diverticula.
12. Two female genital pores on segament 14th.
13. Two pairs of penial setae help in copulation.

ACANTHOBDELLA

Phylum	-	Annelida
Class	-	Hirudinea
Order	-	Acanthobdellida.
Genus	-	*Acanthobdella*

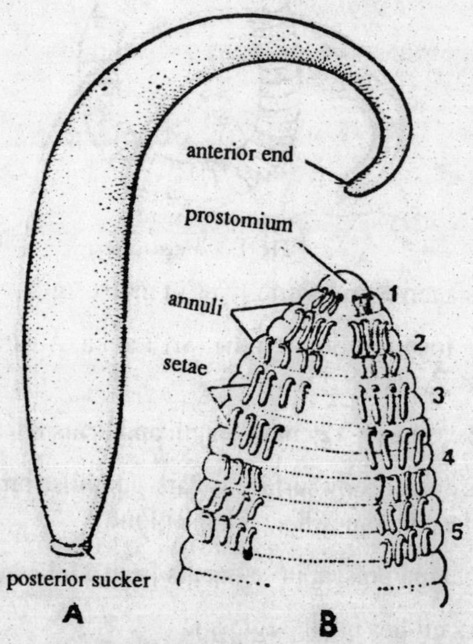

A- Entire Leech

B- Anterior end in ventral view showing setae.

Fig. 1.21 Acanthobdella

1. It is an ectoparasite on the caudal and anal fins of salmon fish.
2. It is found in West Siberia and North-East Europe.
3. Body is elongated dorso-ventrally flatenned and both the ends pointed.
4. It contains only 20 segments.
5. Anterior sucker is absent, posterior sucker made up of four segments.
6. First five segments bear double rows of setae ventro-laterally.
7. Body cavity or coelom spaceous and is incompletely divided by septa.
8. Nephridia with external and internal openings.
9. Circulatory system made up of a dorsal and a ventral blood vessel.
10. Nervous system consists of 20 ventral ganglia.
11. Male genital pore in 7th segment and female genital pore in 8th segment.
12. *Acanthobdella* is a connecting link between Oligochaeta and Hirudinea.

PONTOBDELLA

Phylum	-	Annelida
Class	-	Hirudinea
Order	-	Rhynchobdella
Genus	-	*Pontobdella.*

1. It is an ectoparasite on fishes chiefly skates and rays. Sometimes it is called 'skate sucker'.
2. It is found in Europe and U.S.A.
3. The body is cylindrical, elongated with anterior narrow and posterior blunt portion.
4. The anterior and posteror suckers are cup-shaped and well marked.

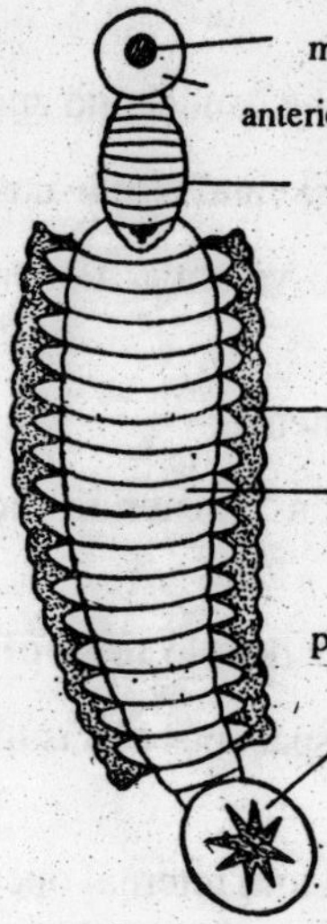

Fig. 1.22. *Branchellion*

5. Anterior sucker is applied to the skin of fish and then makes a wound.

6. The body surface is rough which is covered by small rounded green tubercles.

7. Nephridia form a complex netwrok on the ventral surface of body.

8. The anterior end of body protrudes out to form a retractile proboscis or introvert.

9. Setae and jaws wanting.

10. Hermaphrodite, sexual reproduction is common.

11. Eggs are laid in molluscen shells. Parental care is shown as the eggs are guarded by the parent till they hatch.

BRANCHELLION

The classification is the same as that *Pontobdella.*

1. *Branchellion* is an ectoparasite on marine fishes specially *Torpedo*.

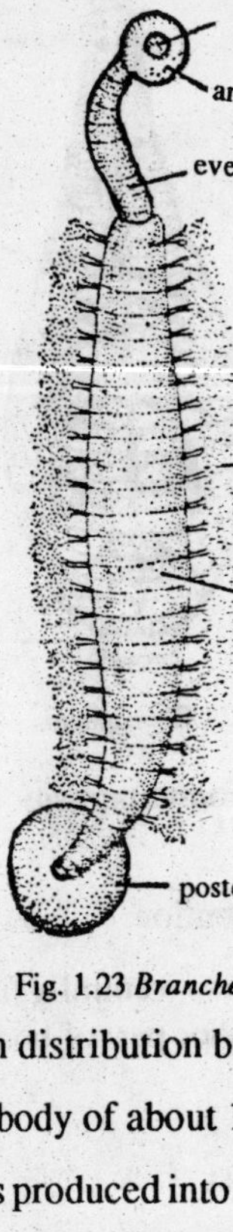

Fig. 1.23 *Branchellion.*

2. It is world wide in distribution but not reported in India.
3. Small, elongated body of about 10 cm. long.
4. The anterior end is produced into a proboscis and a wide abdominal region.
5. The abdomen bears 11 pairs of series of foliaceous non-digitate gills (branchiae) laterally in each segment.
6. The anterior and posterior suckers well developed and pedunculate.
7. Nephridia are absent but excretion by complex network of tubes.
8. Hermaphrodite, sexual reproduction is common.

GLOSSIPHONIA

The classification is the same as that *Pontobdella.*

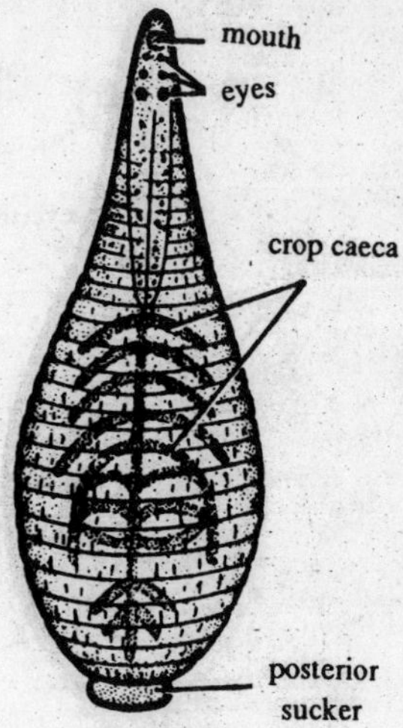

Fig. 1.24 Glossiphonia.

1. It is a freshwater leech found parasitic of muscles and feeds on *Chironomous* larvae.
2. It is world wide in distribution.
2. The body is broad and dorso-ventrally fiattened. The anterior end tapers whereas the posterior part of body is oval.
3. Setae are wanting:
4. Three pairs of eyes persist.
5. The posterior sucker is cup-shaped and well formed.
6. Presence of elongated crop chambers arranged in the middle of body segments.
7. In this leech copulation has never been observed, but one individual deposits spermatophores on any part of the body of another individual.
8. The male and female ducts open by a common pore located between eleventh and twelth segments.
9. Cleavage unequal no cocoon formation takes place.

OZOBRANCHUS

The clasification is the same as that of *Pontobdella*.

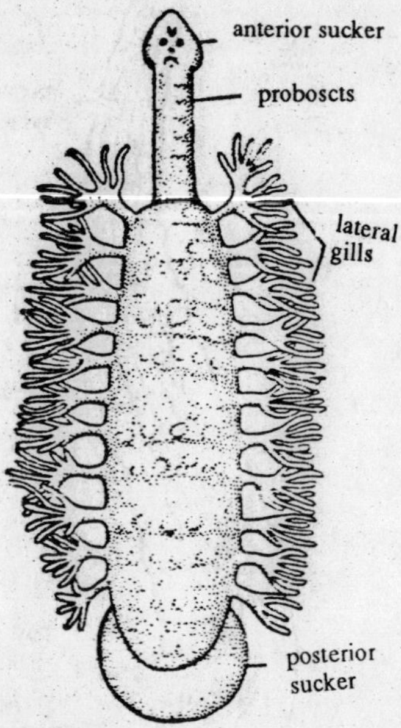

Fig. 1.25 Ozobranchus.

1. It is an ectopara site on the river-turtle and fishes.
2. Body is sub-cylindrical that often bears lateral gills.
3. Usually more than three annuli per segment.
4. Anterior and posterior suckers are well developed.
5. Anterior region is produced into a long proboscis.

HERPOBDELLA (NEPHELES)

Phylum	-	Annelida
Class	-	Hirudinea
Order	-	Gnathobdellida
Genus	-	*Nephelis*

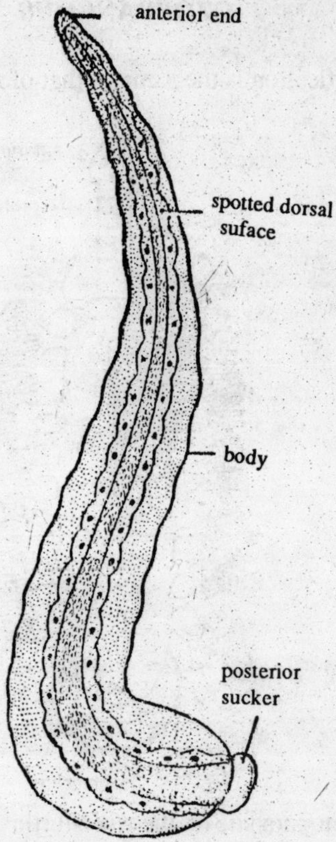

Fig. 1.26 Nepheles.

1. It is found in fresh water ponds and streams under stones and in vegetation.
2. It has world wide distribution.
3. Body is slender about 8 cm in length and brownish black in colour.
4. Anterior end is rounded and devoid of anterior sucker. Head bears 2 pairs of eyes.
5. Each body segment has 5 annuli. The annuli and true segments are not easily demarkated and consists of 19 segments.
6. Posterior end is flatenned and bears a distinct small posterior sucker.

7. Setae, parapodia and jaws are absent.
8. It is carnivorous, feeds on aquatic larvae, worms and snails.
9. Sexual openings are separated from one another by two rings.
10. Development takes place in cocoon.

HAEMOPIS

The classification is the same as that *Nepheles*.

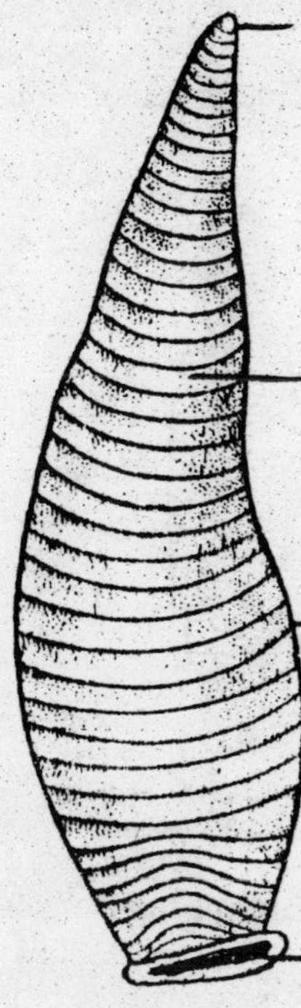

Fig. 1.27 Haemopis

1. It is found in ponds, streams and swampy meadows.
2. It is cosmopolitan in distribution.
3. It is commonly called 'hungry horse leech'.
4. Body segmented, segments indicated externally by ring-like constrictions, which are much more numerous than the true segments.

5. Head is not distinct, jaws are armed with blunt teeth in two rows. A pair of eyes present.
6. Presence of one pair of caeca in crop.
7. It is carnivorous leech, feeding on earthworms, aquatic larvae of insects, tadpoles and small fishes.
8. Copulatory glands are absent.
9. Development takes place in cocoons.

LIMANTIS

The classification is the same as that of *Nepheles*.

Fig. 1.28 *Limnatis*

1. It is a fresh water ectoparasitic leech of domestic animal, horses, cattles etc.
2. Commonly found in Syria, Egypt and Lebanon.
3. Body is elongated never more than 4 mm in length.
4. Anterior and posterior suckers are well developed.
5. Proboscis non-protrusible.
6. Oral sucker armed with three jaws.
7. Each segment with 5 annuli.

8. Development is direct.

POLYGORDIUS

Phylum	-	Annelida
Class	-	Archiannelida
Genus	-	*Polygordius.*

1. It is found living in the sand of European Seas.

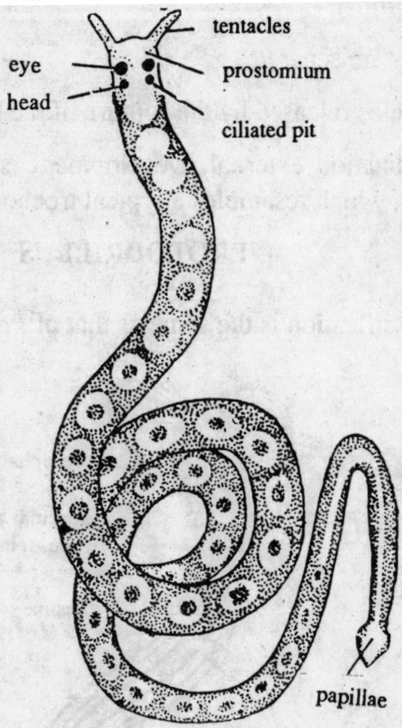

Fig. 1.29 Polygordius

1. *Polygordius* is a marine worm with narrow, elongated cylindrical body.
2. The prostomium is small, the peristomium large and forms the head, which bears paired prostomial tentacles and ciliated pits.
3. Presence of a tooth-like process round the anus, and in front a circlet of adhesive papillae.

4. Segmentation is not distinct in front, but well marked behind and is without parapodia or chaetae.
5. Beneath the skin longirudinal muscle fibres in four quadrants. Circular muscles absent.
6. Presence of paired segmentally arranged nepridia leading into nephrostomes.
7. Vascular system is reduced and single nerve cord lies in the epidermis.
8. Sexes are separate.
9. Gametes released by the rupture of the body wall.
10. Fertilization external. Development is indirect with Loven's larva, which resembles a typical trochophore larva.

PROTODRILLUS

The classification is the same as that of *Polygordius*.

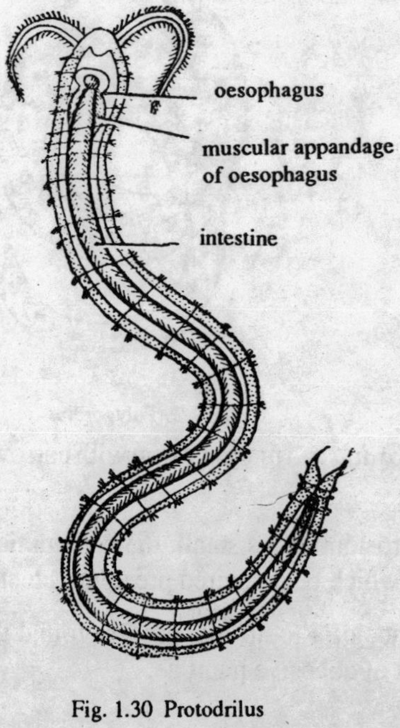

Fig. 1.30 Protodrilus

1. It is small marine worm found in Inland Sea, Lake at Faro near Messina (U.S.A.).
2. Body is narrow elongated and cylindrical.
3. Body segments are very indistinct and marked off by ciliated rings.
4. There is a median longitudinal groove on the ventral surface.
5. Body with small prostomium which bears cephelic tentacles and ciliated pits.
6. Anus is terminal surrounded by fin-like expansions.
7. The ventral nerve cords are fused.
8. Excretory system comprises paired segmental simple nephridia.
9. Hermaphrodite. Development is direct without larval stage and metamorphosis.

SACCOCIRRUS

The classification is the same as that of *Polygordius*.

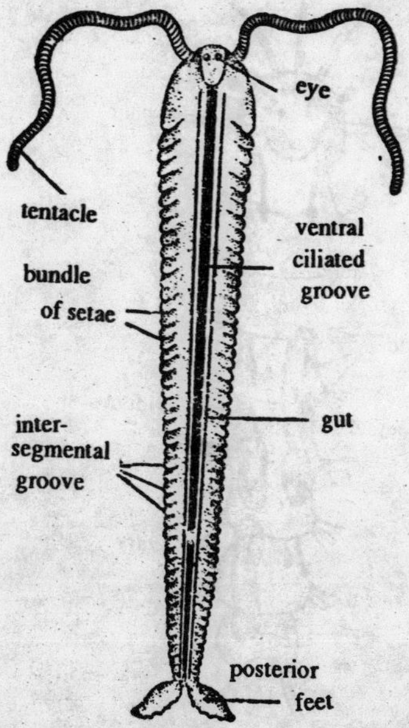

Fig. 1.31 Saccocirrus.

1. It is marine worm found attached to sea-weeds by adhessive pad.
2. It is found in European countries and Madras.
3. Body long, cylindrical, 10-15 mm in lingth.
4. Anterior region bears a pair of cephalic tentacles and eyes.
5. Body segments marked by prominant inter-segmental grooves.
6. Each segment bears lateral setae.
7. Longitudinal ciliated groove is situated in the middle of ventral surface.
8. Anus terminal with adhesive pads on.
9. Nervous system comprises of two separate ventral nerve cords lying just beneath the epidermis.
10. Sexes separate, female with spemathecae and males with a pair of protrusible penis in each segment.
11. Development is direct.

NERILLA

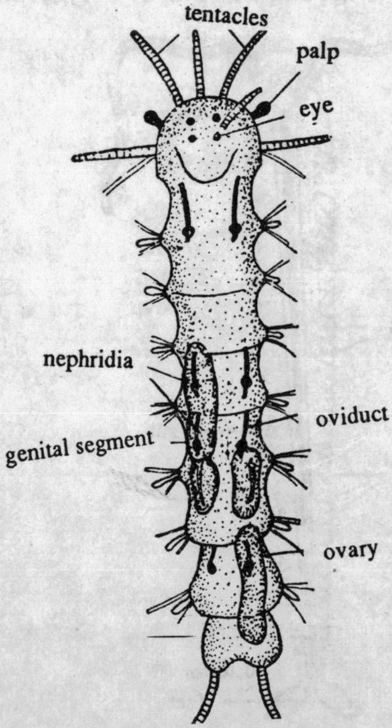

Fig. 1.32 Nerilla

The classification is the same as that of *Polygordius*.

1. It is a marine worm found in sea-water.
2. Body elongated worm-like measuring 1 mm in length.
3. Head with three prostomial tentacles, two palps and two pairs of setae.
4. Sexes are separate.
5. Three genital segments in male and one in female.
6. In male three pairs of sperm ducts uniting at a common median genital aperture.
7. Development is direct.

DINOPHILUS

The classification is the same as that of *Polygordius*.

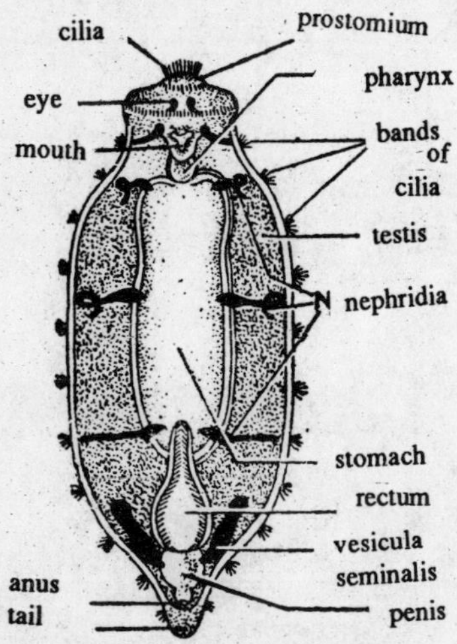

Fig. 1.33 Dinophilus.

1. *Dinophilus* is small, marine archiannelid, found among sea weeds.
2. Body short, flattened looks like a planarian of about 2 mm long.
3. Body with 5 to 6 segments with ciliated ring.

4. The entire ventral surface is uniformly clothed with cilia.
5. Prostomium broad, with two eyes and ciliated pits.
6. Tentacles and palps are absent.
7. Coelom very much reduced with five pairs of nephridia. Sexes are separate.
8. Male with median penis.
9. Females with eggs of two sizes the smaller giving rise to males and the larger to females.

2

NEANTHES = NEREIS

Neanthes (old name *Nereis*) is commonly known as *sandworm* because it is found buried in sand or *clamworm* as it is found along with clams but has no ecological relatioship with them. *Neanthes* is one of the largest and most common typical marine annelid with most of the characteristics of the phylum. It has a great number of species which are found in different parts of the world but resemble greatly with each other. Few important and common species are : *N. dumerilli, N. diversicolar, N. irrorata, N. cultrifera, N. virens, N. pelagica,* and *N. caudata.*

SYSTEMATIC POSITION

Phylum	...	Annelida
Class	...	Polychaeta
Order	...	Errantia
Genus	...	*Neanthes* (*Nereis*)

Habits and habitat

Neanthes has world wide distribution i.e. it is found throughout the world, in Europe, America, Alasks, North Pacific and other places. It inhibits the sea-shores between tide marks. It is also found among the sea-weeds and under stones, but normally it lives in burrows. The burrows are formed 60 cm deep. The burrow is lined by muscous which help in cementing the grains of sand, a mechanism by which the burrow cannot collapse. The worm lives within the burrow during day time. At night, it protrudes the anterior of its body and moves it all round in the search of food. Thus the worm is nocturnal. The worm is carnivorous in diet and feeds on small crustaceans, molluscs and annelids. Occasionally, the worm leave its burrow and creeps about the stones. During breeding season, it leaves its burrow permanently and swims about actively in the surface layers of water.

External morphology

Shape and size.-Body is elongated, cylindrical and vermiform. It is *bilaterally symmetrical, metamerically segmented* and somewhat flattened dorsoventrally. It gradually tapers to the posterior end. The dorsal surface is convex but the ventral flat. The size varies from few to 40 cm.

Colouration. Different species are differently coloured. *N. virus* is steel-blue in colour; *N. limbata* is of brownish colour while jaws are light amber - coloured; *N. cultrifera* is greenish; *N. pelagica* is raddish-brown; *N. lamnicola* is also raddish-brown; and *N. vexilosa* is dark brown or blue-green with iridescence.

Segmentation-Body is divided into approximately 200 or more *segments* or *metameres.* In *N. cultrifera* there are 80 segments. These are marked externally by grooves and internally by septa. All the metameres are alike except the last which is somewhat rounded. It is known as *tailsomite* or *pygidium* and carries a pair of *anal cirri.* Anus is terminal.

Divisions of the body. The whole body of *Neanthes* is divided into three well-defined regions.

1. Head
2. Trunk
3. Pygidiuum.

1. Head.

The anterior-most part of the body is specialized into a head, which consists of two main parts, *prostomium* and *peristomium.*

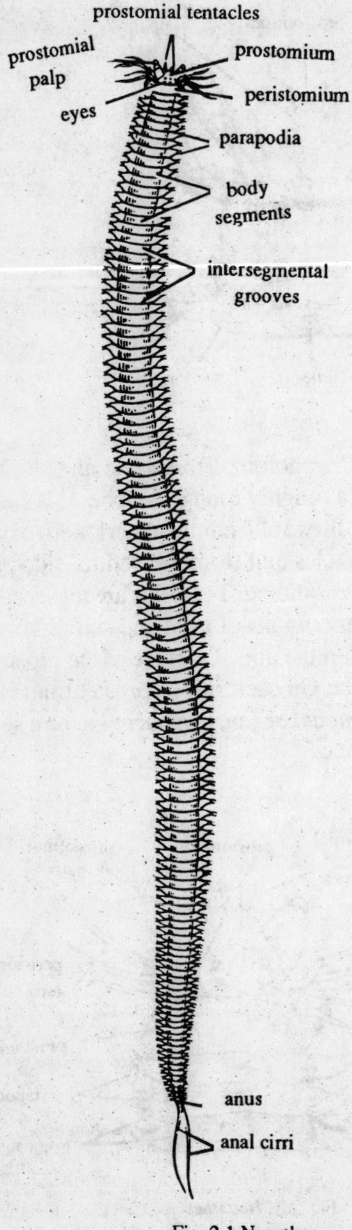

Fig 2.1 Neanthes

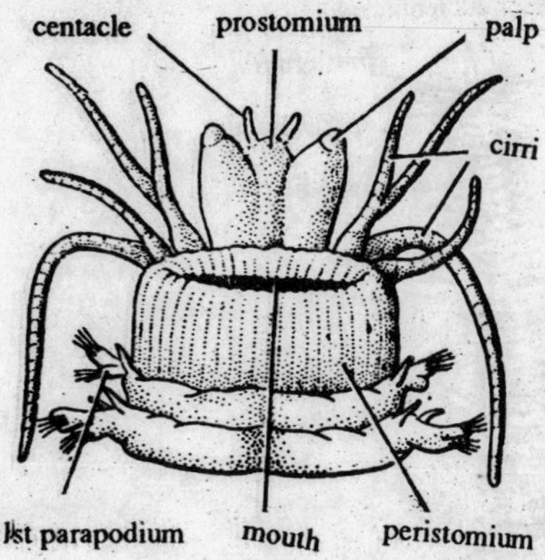

Fig 2.2 *Neanthes*

(i) *Prostomium* (Gr., *Pro*=before; *stoma*=mouth). It is the part before mouth and is a roughly triangular lobe. It is situated mid-dorsally. Protruding forward from the anterior edge of the prostomium are two tentacles and from the ventral side project two *palpi*. Each palp is two-jointed. Tentacles are regarded as organs of touch, and palpi are organs of taste and smell. There are two pairs of simple eyes on the dorsal surface of the prostomium for the perception of light. On each side of prostomium is a ciliated pit, which is called *nuchal organ*, considered to be a sense-organ of doubtful importance.

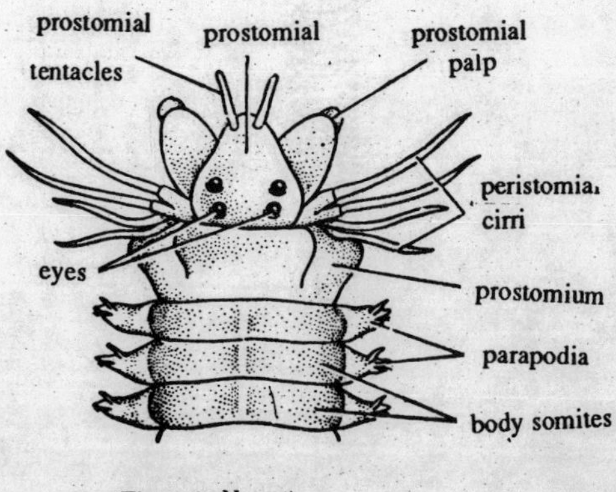

Fig 2.3 *Neanthes*

(ii) *Peristomiuum* (Gr. *peri*=around ; *stoma*=mouth). It is a large ring-like round part present behind the prostomium. It bears on its ventral side a slit-like *mouth*. It differs from a trunk segment in being longer, in lacking *parapodia* and in the presence of four (two pairs) thread-like *peristomial cirri* on each side; of these one pair is dorso-lateral and the other ventro-lateral. These are homologous with the notopodial cirri and neuropodial cirri of the parapodia of trunk segments. Each cirrus is a long, slender, tactile structure, having a short proximal joint and a long distal joint. The dorsal cirri are longer than the ventral ones.

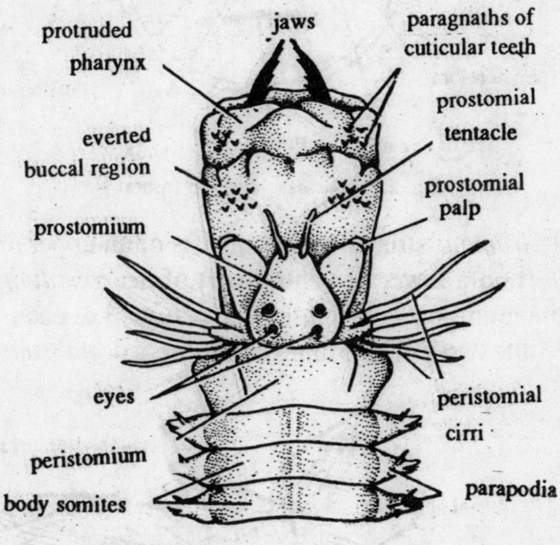

Fig 2.4 *Neanthes*. Head in dorsed view with everted pharynx,

The study of the developmental stages of *Neanthes* shows that the prostomium is not a true segment of the body but develops as a dorso-anterior projection of the peristomium, whereas the peristomium is formed by the fusion of the first and second body segments.

2. Trunk

It comprises practically the entire body except the head and the last segment or pygidium. It comprises a number of segments which are all alike. Each trunk segment bears on either lateral sides a flat, fleshy, hollow and vertical flap-like structure, the parapodia. Term parapodia was coined by *Huxley* (Gr. *para*=besides; *podos*=foot)

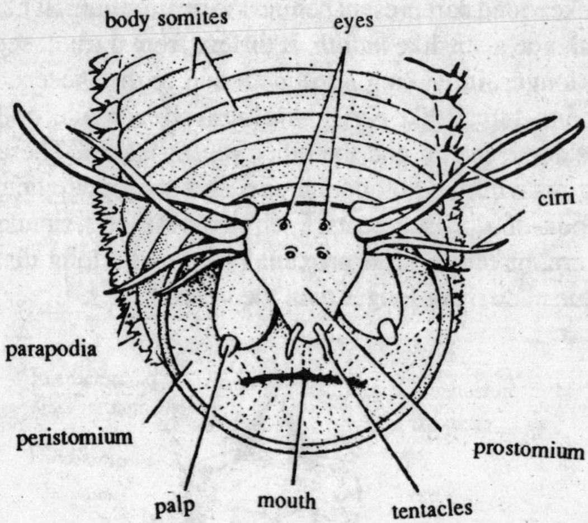

Fig 2.5 *Neanthes*. Head in frontal view.

It is a *biramous* structure and consists of an upper or dorsal part or *notopodium* and a lower or ventral part of *neuropodium*. Notopodium and neuropodium are both bilobed structures i.e. each part is further subdivided into two leaf-like lobes or *ligulae*, a dorsal *superior ligula* and

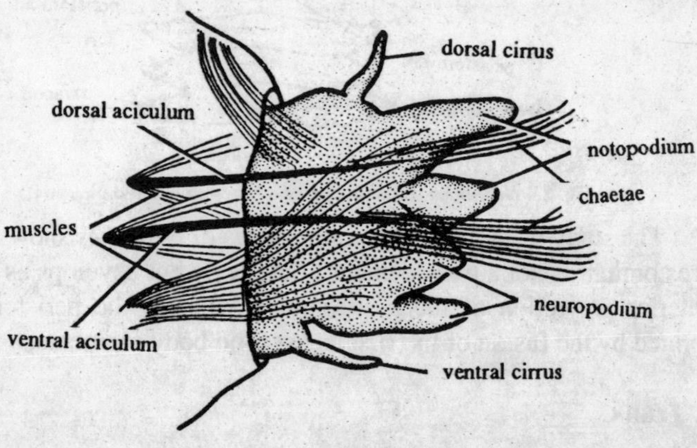

Fig 2.6. *Neanthes*. Parapodium.

a ventral *inferior ligula* and from each apperas a bundle of setae or chaetae. The setae lodged in an epidermel pit, the *setigerous sac*. Each seta develops from a single large formative cell, present at the base of setigerous sac. The setae can be protruded, rotated in all the directions and retracted by special muscles within the parapodia. Each set is two jointed structure with aproximal *shaft* and a distal *blade*. The setae are of two types (a) Typical type has a long stout shaft and a short stout blade, (b) Long blade type has a small shaft but a long, pointed blade with one edge serrated. In Heteronereis phase a third type is also seen called as oar-shaped. Here the blade becomes ora-shaped. Besides long setae in each part, there is a stouter, straight, dark-coloured seta known as *aciculum*, which projects only for a short distance but not beyond the outer edge of the parapodium. Setae help in locomotion, while *aciculum* has a skeletal function. The dorsal margin of notopodium is produced into a short, cylindrical tactile appendage, the *dorsal cirrus* and a similar structure is produced at the ventral margin of the neuropodium, the *ventral cirrus*. Of the two cirri, the dorsal one is much larger than the ventral.

The parapodia are the largest in the mid-region of the body and decrease in size towards the anterior and posterior ends of the body. The first two parts of parapodia have no notopodial setae. The parapodia are highly muscular and primarily the organs of locomotion, used both in creeping and in swimming. They are highly vascular and glandular and probably also serve as respiratory organs.

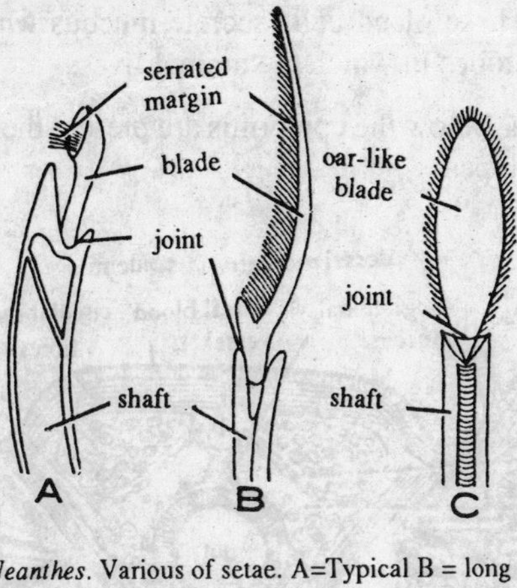

Fig 2.7 *Neanthes*. Various of setae. A=Typical B = long blade, C=Oar shaped

Nephridispores. On the ventral side, near the base of parapodia, each segment bears a pair of minute excretory pores through which the nephridia communicate out.

3. Pygidium or Tail.

The last segment is tail which bears a terminal anus, a pair of long filamentous *anal cirri* and several minute sensory papillae. This segment does not bear parapodia.

Body wall.

The body wall consists of cuticle, epidermis, musculature and coelomic epithelium.

1. *Cuticle.* It is a thin, tough, chitinous and non-cellular layer covering the body externally. It is secreted by the epidermal cells underlying it. It is pertorated by numerous pores through which the epidermal gland cells open out. The cuticle exhibits an irridescent lusture due to the presence of two intersecting systems of fine striations in it.
2. *Elidermis.* It lies below the cuticle. It is made up of a single layer of columnar supporting cells resting on a basement membrane. Except the ventral side, the epidermis is thick at places. Many gland cells are found on the ventral surface near the base of parapodia. These gland cells secrete mucous which is used in forming the tubes in which the animal lives.
3. *Musculature.* Below the epidermis are present the musles which are of three types:

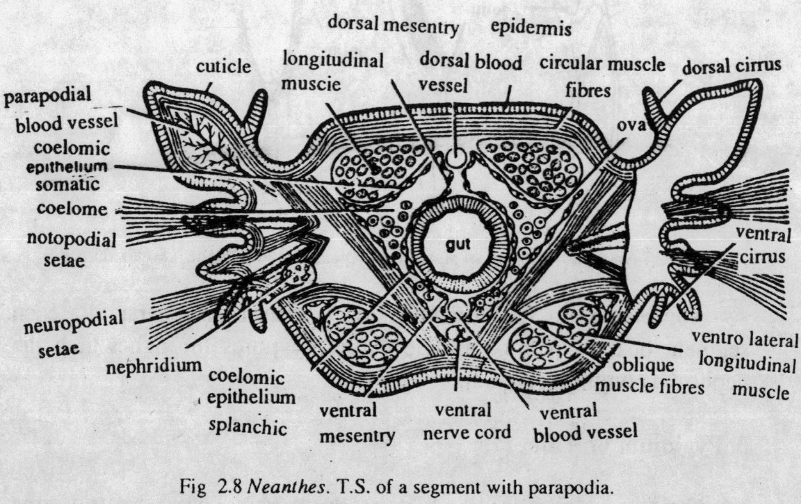

Fig 2.8 *Neanthes*. T.S. of a segment with parapodia.

(a) Circular, (b) longituudinal and (c) oblique muscles.

(a) *Circular muscles.* These form a continuous layer beneath the epidermis, which is slightly thicker on the ventral side. In the parapodia, the circular muscles are modified to form a complicated system of parapodial muscles comprising the *protractor* and *retractor* muscles. The protractors radiate from the bases of the setigerious sacs to the layer of circular muscles on all sides. The retractors extend from the outer part of the setigerious sacs to the dorso-lateral body wall. By the contraction of protractor muscles protrusion of acicula and central lobes of parapodia occur. The retractor muscles brings about withdrawl of the acicula-bearing lobes of the parapodia.

(b) *Longtidinal muscles.* The longitudinal muscles are inside the circular muscles and are much better developed. They do not form a complete layer but occur as four bundles, two dorso-lateral and two ventrolateral. Contraction of this muscle makes the body short and stout.

(c) *Oblique muscles.* Two pairs of oblique muscles are found in each segment. They arise from the ventral body wall on the lateral sides of the never cord extending forwards to attach with the bases of the parapodia. The oblique muscles are divided into two parts, one part is attached to the dorsal part of the base of the parapodia and the other to the ventral part. Bending of parapodia in the vertical and horizontal planes is brought about by differential contraction of the various oblique muscle-bundles.

All the nuscles are composed of smooth or unstriped muscles fibres.

4. *Coelomic epithelium.* It is the inner most layer of body wall and is composed of single layer of thin cells.

Functions of body wall. The body wall performs a number of functions which are as follows:

(i) The cuticle protects the body from mechanical injuries and dessication.

(ii) The epidermal gland cells secrete mucous which lines the burrows to prevent the collapse.

(iii) The epidermis of dorsal side is highly vascular which helps in respiration.

(iv) The sensory cells make the animal sensitive to external stinuli.

(v) Muscles help in movements and serve for locomotion.

(vi) The coelomic epithelium secretes coelomic fluid.

Coelom

Between the body wall and gut intervenes a fairly specious *coelom* or *perivisceral cavity*. This is not, however, a continuous cavity but, on the contrary, is broken into a series of coelomic chambers by the intervention of intersegmental septa, corresponding to the external groove. However, the septa are incomplete, especially round the alimentary canal, so the contiguous coelomic chambers communicate with each other. The coelom is further partitioned by the presence of a pair of oblique muscles running obliquely dorsoventrally from the ventral to the dorsolateral body wall. Ventrally, these museles are inserted into the body wall below the ventral nerve cord; dorsally they are attached to the body wall near the bases of parapodia. The perivisceral cavity is lined by coelomic epithelium or peritoneum. Peritoneum that lines the inner surface of body wall is called *parietal peritoneum* and that covers the outer surface of viscera is called *visceral peritoneum.* A double fold of peritoneum called *mesentery* suspends the alimentary canal into the coelom from mid-dorsal body wall.

The coelom is filled with *coelomic fluid.* This fluid contains amoeboid corpuscles. The coelomic fluid more or less freely circulate throughout the body.

Functions of coelom. The coelom performs following functions:

1. Food and oxygen are supplied to different parts of the body with the help of coelomic fluid.
2. It protects the body from external shocks.
3. The coelomic fluid also helps in the protrusion of parapodia during locomotion.

Locomotion

Two structural elements are especially important in the locomotion of *Neanthes*, the longitudinal muscles and parapodia. The longitudinal muscles do not form a continuous layer. Instead they are broken up into two pairs of blocks, one pair dorsal and other ventral. Because of this the muscles of the two sides of a segment can be in opposite phases, one contracted and other relaxed so that the passage of waves of contractions along the body can be thrown it into lateral undulations. Assisted by the parapodia the lateral undulations displace the body forward.

The parapodium in *Neanthes* is clearly quite versatile in its mode of action. The animal remains dependent upon the hydrostatic skeleton to provide firm basis for muscular action. The properties of musculature and parapodia interact to provide for several different types of movements. These movements are of three types :

1. Slow creeping
2. Fast creeping
3. Swimming

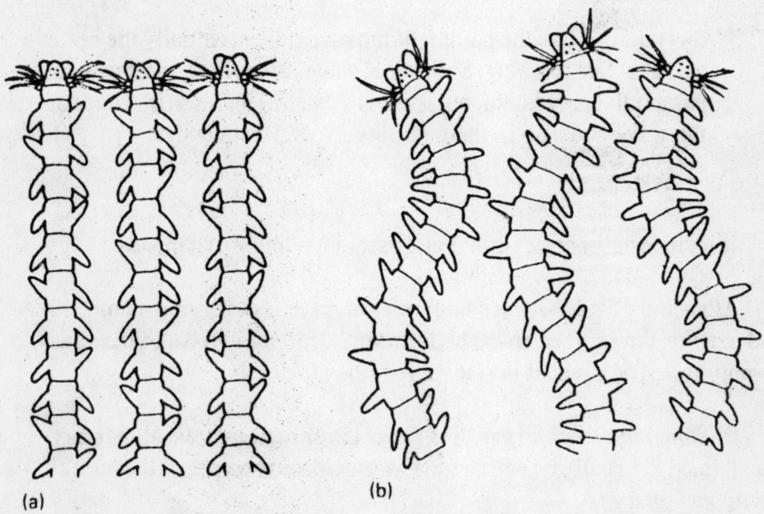

Fig 2.9 Diagrammatic representation of locomotion in *Neanthes*.
(A) Slow crawling (B) = rapid crawling.

1. *Slow creeping*. Slow creeping depends almost entirely upon the use of parapodia as a series of levers. Initially a parapodium moves forwards, with its tips lifted from the ground and with the

aciculum withdrawn, this is called the *recovery stroke* or *preparatory stroke*. At the end of this stroke the parapodium makes contact with the substratum, the aciculum is protruded, and the oblique muscles contract so that the body is pulled forwards. The parapodium is brought to backwards during this phase which is called *effective stroke* or *power stroke*. The movements of parapodia are so integrated that the two members of any one segmental pair alternate with each other in phase. Moreover, the movement of any one parapodium begins slightly after that of the one next behind it.

2. *Fast creeping*. Slow creeping readily passes into rapid creeping. This involves a similar rhythmic pattern of waves, but with the difference that longitudinal muscles are now of primary importance. They contract serially in parallel with the movement of parapodia. The oblique muscles are now probably of much less importance than in show creeping.

3. *Swimming*. Here the pattern of movement is essentially the same as that of fast creeping. Sinusoidal waves of the body interact with parapodial movement, but there is a marked increase in length of the wave and also in their amplitudes and frequency.

Digestive System

It includes alimentary canal and associated digestive glands.

Alimentary Canal. It is a complete straight tube of varying diameters and extends through the entire body from the mouth in front to the anus behind. It can be divided into three regions :

(i) *Stomodaeum* or *forgut*: It is the anterior region of the alimentary canal, lined internally by ectoderm and cuticle and comprises the buccal cavity and pharynx.

(ii) *Mesenteron* or *midgut*: It is the middle region of the alimentary canal and it is lined internally by endoderm and comprising the oesophagus and stomach-intestine.

(iii) *Proctodaeum* or *hindgut*: It is the posterior-most part of the alimentary canal, lined internally by ectoderm and cuticle and comprising only rectum.

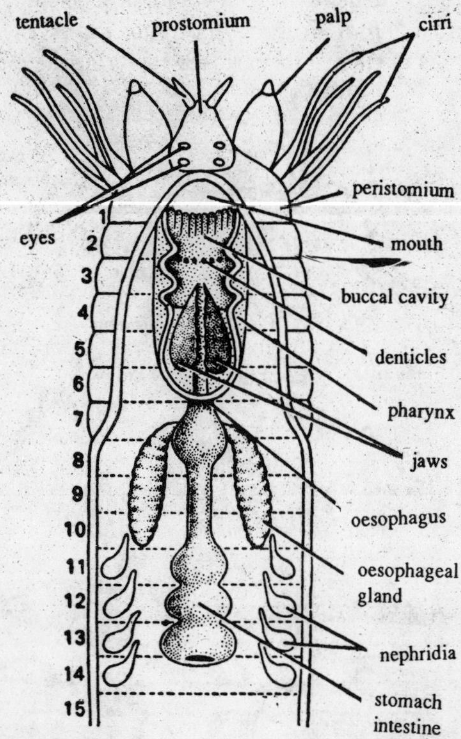

Fig 2.10 *Neanthes*. Anterior end dissected dorsally to show alimentary canal

(a) *Mouth*. It is a transverse slit on the ventral side of peristomium. It is over hung by the prostomium. It leads behind into buccal cavity.

(b) *Buccal cavity and pharynx*. The anterior part of the buccopharyngeal regions of *Neanthes* forms an eversible *proboscis*. It is sorrounded by several coasts of nuscles. The buccal cavity is situated in the peristomium and leads behind into the pharynx which extends upto 4th segment. Both of them are lined internally by cuticle. The cuticle of buccal cavity is thckened at various points into dark brown *denticles* or *paragnaths*. The pharynx is thick-walled and muscular. In its posterior part are found a pair of large, powerful, chitinous and movable jaws. Each jaw has a hollow *base* and an incurved, pointed, somewhat notched *apex*. The inner margins of the jaws are serrated. The buccal cavity and pharynx are wrapped in a common muscular coat and together they extend up to the fourth or fifth trunk segment. They can be fully everted to form the *proboscis* or the *introvert*. This operation exposes the tips of the jaws for capturing the prey.

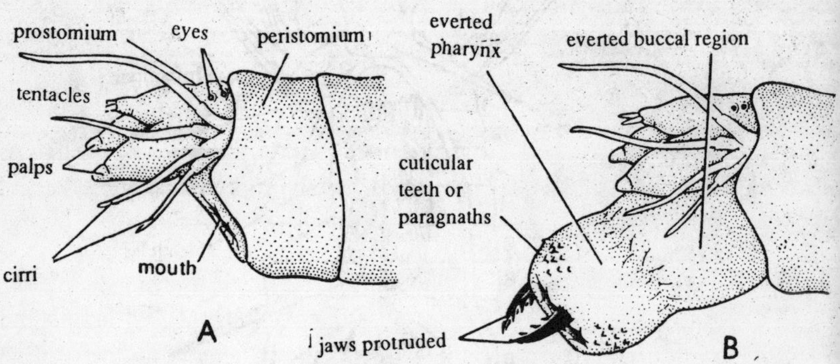

Fig 2.11. *Neanthes* A = Pharynx retracted B= pharynx everted.

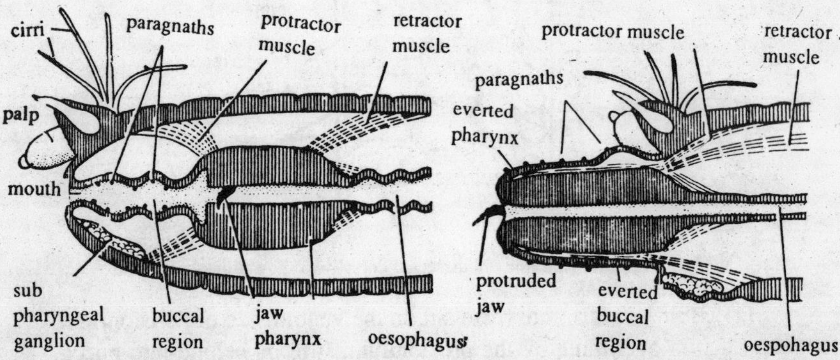

Fig 2.12 *Neanthes* Diagramatic representation of the action of introvest

(A) = at rest (B) = protruded introvest

(c) *Oesophagus.* The pharynx narrows posteriorly to lead into the oesophagus. It is a narrow tube extending through five segments behind the pharynx. A pair of long, unbranched, sacculated glandular puches, the *oesophageal glands* or *caeca*, open into it laterally at the anterior end. The oesophagus opens into stomach-intestine; the opening between the two is guarded by a sphincter muscle. A distinct stomach is absent.

(d) *Stomach-intestine.* It is a wide tube extending through the remaining length of the body. It is straight, thin-walled tube and segmentally constricted by septa. The stomach-intestine is the principal site of digestion and absorption. The stomach-intestine opens into a rectum which lies in the last segment.

(e) *Rectum.* It is the last part of alimentary canal, which opens posteriorly by the anus. It is very short chamber and occupies the last body segment or pygidium.

Digestive glands. As stated above into the oesophagus open a pair of laterally placed caecal or poches called *oesophageal glands.* They secrete digestive juices probably proteolytic in nature. Besides microscopic gland calls are present in stomach-intestine. These glands secrete digestive juices that contain euzymes to digest food.

Histology. The gut wall consists of an outer most layer of serosa or visceral peritoreum, followed by longitudinal muscles, circular muscles and an enteric epithelium which is endoderm in the case of mid-gut and ectoderm in the case of fore-gut and hind-gut. The ectoderm is lined by cuticle.

Food. Neanthes is a *carnivorous* and *voracious* animal and feeds on small animals, such as molluscs, crustaceans, sponges and other. It is also called clam-worm, because it is generally found, where clams occur. But it never eats the living clams as they are of bigger size. However, it feeds on small animals.

Feeding. To capture the prey, *Neanthes* everts its stomodaecum. The *eversion* of the pharynx is brought about by the contraction of *protractor muscles* extending from the introvert to the wall of peristomium and also by the pressure of coelomic fluid. The muscles contract and the coelomic fluid gets filled in the buccal region, so that the buccal cavity everts out like a bag and the pharynx protrudes out. The two pharyngeal jaws are thrust with their dentate margin facing outwards. The jaws come close together and grasp the prey. The food is now dragged inside the burrow and then swallowed by the withdrawing of pharynx. The retraction of introvert is effected by contraction of retractor muscles which extend from the body wall to the pharynx.

Filter feeding. When within its burrow, *Neanthes* feeds on minute organic particles brought into the water current with the water current.

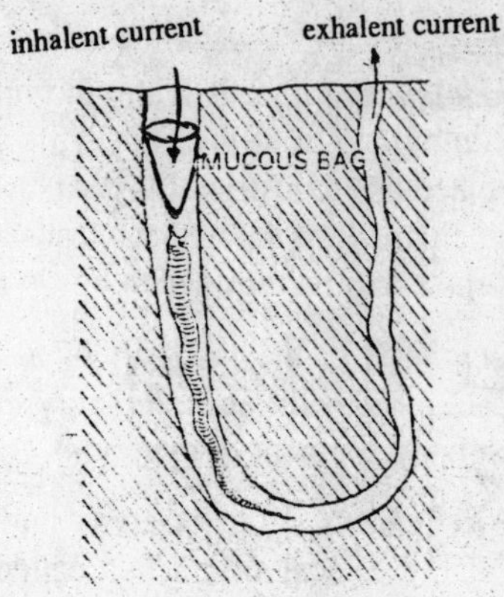

Fig 2.13 *Neanthes*. Filte feeding in burrow.

The water current is created by lateral undulation of the body. For this purpose it secretes a mucous bag. Particles collected in the secretion, as thought in a bag, are swallowed from time to time. This mode of taking food is called filter feeding.

Digestion and absorption. Digestion is intercellular, occurs in the lumen of mid-gut. It is brought by the enzymes secreted by oesophageal glands and gland cells of stomach-intestine. The digestive enzymes are proteases, amylases and lipases. Absorption mainly occurs in the stomach-intestine.

Respiratory System.

In *Neanthes*, respiratory organs are absent. Respiration occurs by diffusion through body surface. The parapodia with their rich-blood supply and body wall with its plexuses of blood vessels subserve the function of respiration. When the blood passes through these net-works, it obtains oxygen from the surrounding water and gives out carbon dioxide. Blood contains a respiratory pigment *heamoglobin*, which increases the absorptive capacity of blood for oxygen and carbon dioxide.

Circulatory System.

In *Neanthes* a closed type of circulatory system is present which helps in transportation of materials from one region to other region of body. The function of transportating materials is performed by the blood. The blood flows in a system of tubes, the blood vessels. The blood vessels are of two types distributing and collecting. Both the vessels are connectged to each other by a system of extremely fine vessels, the *capillaries*. By the peristaltic contraction, the blood is maintained in constant circulation.

Blood vessels. There are three longitudinal vessals which are as follows:

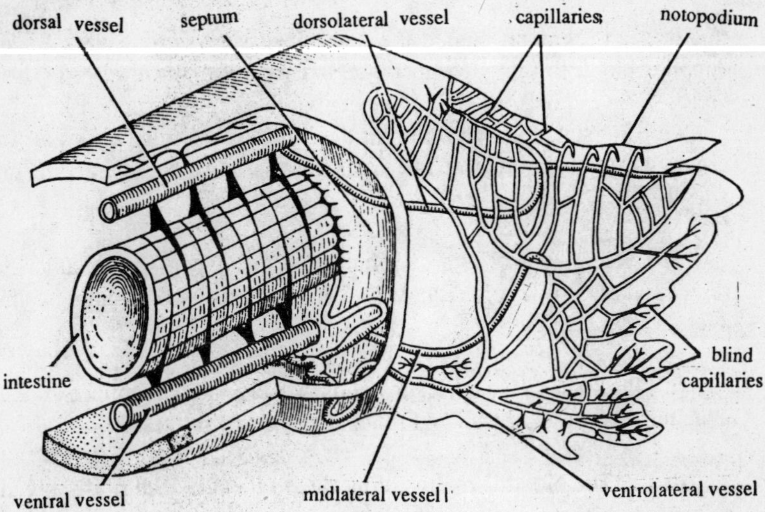

Fig 2.14 *Neanthes*. Diagrammatic representation of blood vessels in T.S. of a segment

(i) *Dorsal vessel.* It runs from one end of the body to the other situated in the dorsal mesentery above the alimentary canal. Its walls are highly contractile which drive blood from posterior to the anterior end. It acts as a collecting vessel, but in the 5th segment it bifurcates and both the branches form a plexus to supply the oesophageal wall. Thus dorsal vessel collects blood from the stomach-intestine region by means of two pairs of *dorso-intestinal* or *efferent intestinal vessels* in each segment. It also collects blood from the body wall, parapodia and nephridia through lateral vessels.

(ii) *Ventral vessel.* It is main distributing vessel. It is situated below the gut and in it the blood flows anteriorly backwards. In the last segment it communicates with the dorsal vessel by a simple *circum-rectal ring.*

In each segment,behind the oesophagus, the ventral vessel is connected with the dorsal vessel by a pair of loop-like *lateral commissural vessels*. Each of these vessels shortly divides into *afferent* branches, which carry blood to the body wall by *afferent cutaneous*, nephridium by *afferent nephridial* and parapodium by *afferent parapodial* of its side. In these organs, the afferent branches break up into a network of capillariess which reunite to form the corresponding efferent branches. The efferent branches finally unite and open into the dorsal vessel.

In each segment, behind the oesophagus, the ventral vessel also supplies blood to the gut wall through a pair of *afferent intestinal* or *ventro-intestinal vessels*. These break up into a capillary network in the gut wall. As stated above from gut the blood is collected by a pair of dorso-intestinal or efferent intestinal vessels and poured into dorsal vessel.

The ventral vessel is a distributing vessel except in the anterior region, where it is a collecting vessel and collects blood from oesophageal wall.

(iii) *Peri-neural vessel*. It surrounds the nerve cord. It collects blood from the ventral body wall and sends it to the ventral vessel.

Blood. The blood is red in colour due to the presence of dissolved respiratory pigment, the haemoglobin. In the blood are present blood corpuscles. The blood corpuscles are colourless, nucleated, amoeboid and resemble the leucocytes of vertebrate blood.

Blood performs a number of functions. It helps in transporting materials, removal of waste products, removing of microorganisms and respiration i.e. exchange of gases.

Excretory system.

Excretory system consists of metamerically or segmentally arranged pairs of coiled and ciliated tubes, called *segmental organs* or *nephridia*. Each segment, except the peristomium and the pygidium, possesses a pair of nephridia.

Each nephridium is made up of a syncytial mass of protoplasm containing many nuclei. Each nephridium is divisible into two parts:

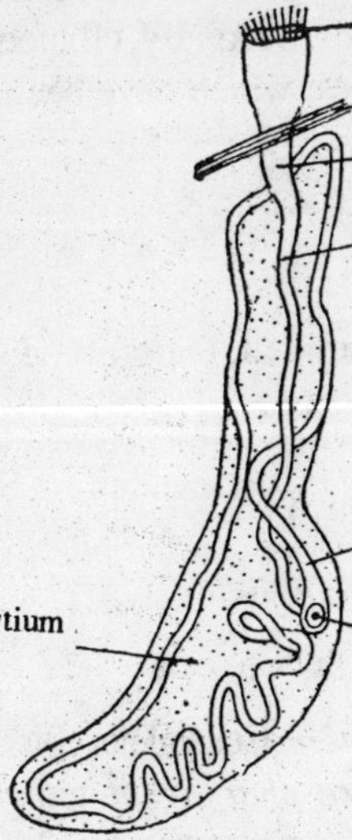

Fig 2.15 *Neanthes*. A nephridium

1. Body of nephridium

2. Neck of nephridium

1. *Body of nephridium.* It is a roughly oval, compact gland shaped structure which is placed transversely in each segment and contains a coiled ciliated tube in it. The cilia are not found in the its last part called *terminal duct*. The terminal duct opens to the outside by a minute aperture, the *nephridiopore*. The nephridiopore is situated on the ventral surface of parapodia near the ventral cirrus. Nephridiospore can be opened or closed.

2. *Neck of nephridium.* The distal part of nephridium is called the neck. In the neck the wall of the ciliated tube is expanded. This part crosses the septum of segment to open into the coelomic cavity of the other where it forms a ciliated funnel-like structure called *nephrostome*. The wide margin of nephrostome is provided with long delicate cilia. Such an open type of nephridium with a ciliated nephrostome is called a *metanephridium*.

In each segment of the body close to the dorssal wall, there is found a pair of *ciliated organs*. These are considered to be the gonoducts of mesodermal origin, which, however, have lost their external openings and have ceased to function as such.

Excretion is possible by two methods. The outer surface of each nephridium is densely covered with blood capillaries. The cells forming the ciliated tube collect the waste materials from the blood and eleminate it through the nephridiopores. Secondly, ciliated funnel helps in removal of dead coelomic corpuscles, which have been perished by bacteria or some foreign bodies.

Nervous System.

Neanthes is an active animal, therefore, a well developed nervous system is present. It consists of :

(i) Central nervous system

(ii) Peripheral nervous system

(iii) Visceral nervous system

(i) *Central Nervous System.* Central nervous system consists of brain or *cerebral* or *supra-pharyngeal* ganglion, which is a large bilobed mass of nervous tissues sitated in the prostomium above the pharynx in the head. This is connected to a *subpharyngeal ganglion*, lying ventrally to the pharynx in the 3rd segment by a nerve-collar or *circumoesophageal connective* or commissure on either side that forms a *ring* or a *collar* around the pharynx. Following the subphyaryngeal ganglion is the ventral nerve, which is ganglionated and is situated in the midventral line throughout tthe complete length of the body below the ventral vessels. In *Neanthes*, two ventral nerve-cords are not separate but are fused together along their inner sides and are enclosed in a common connective tissue sheath. The nerve cord is provided with a ganglionic enlargement in each segment. The first ganglion is situated in the 4th segment.

(ii) *Peripheral nervous system.* The peripheral nervous system mainly includes the nerves coming directly from the brain and nerve cord to supply the different regions of the body. The brain gives out four short optic nerves to the eyes, two tentacular nerves to the prostomial tentacles, and two palpal nerves to the palps.

The ganglia of the circumpharyngeal connectives supply two pairs of *tentacular nerves* to the two ventro-lateral pairs of

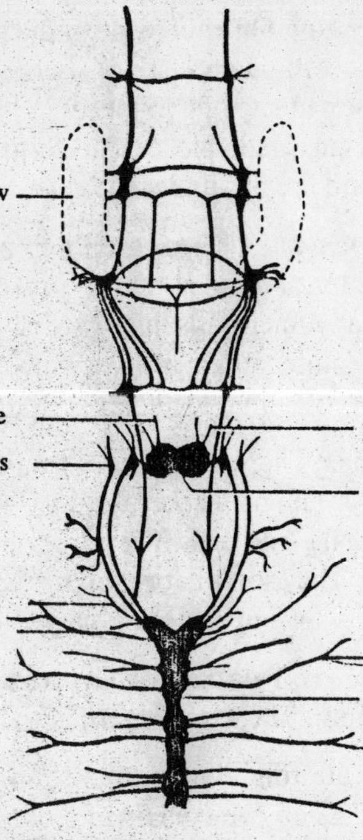

Fig 2.16. *Neanthes*. Nervous System.

peristomial tentacles. The ganglia of the accessory connective innervate the two dorso-letral pairs of prostomial tentacles by a pair of tentacular neves. From the subpharyngeal ganglion a pair of nerves goes to the first parapodia. three pairs of nerves airse from each ganglion of the nerve cord, the first goes to the anterior segment and the remaining two pairs innervate the organs of the segment, parapodia and body wall.

(iii) *Visceral nervous system.* It indudes ganglia sitated above the pharynx and a few fine nerves. These ganglia and nerves are connected on one side with the brain and on the other side with the ventral surface of the circumpharyngeal connectives.

Sense organs.

In *Neanthes* the sense organs are specialized and well developed. Following types of sense organs are present :

1. *Prostomial palps.* On either side of the prostomium from ventro-leteral surface originate a pair of short, thick and muscular palps. Each palp is made up of two pieces; a basal piece, which is large and a terminal small piece. The terminal piece is capable of retraction into the basal piece. These are tectile in function.

2. *Prostomial tentacles.* These are a pair of small, cylindrical arise from the anterior border of the prostomium. The surface of tentacles bear numerious sensory spiral ogans, each consisting of about 100 photoreceptive cells, spirally arranged within a cuticular pit. The prostomial tentacles are probably tectile.

3. *Nuchal organs.* These are a pair of small pits in the posterior past of the prostomium on its dorsal side. They are lined with ciliated epithelium and gland cells. They are connected with the posterior part of brain. They serve as organs of smell and chaemo-reception and help the worm in detecting prey.

4. *Peristomial cirri.* Two pairs of cirri are situated at the anterior end of peristomium. These are regarded to be tectile organs.

5. *Eyes.* There are four simple eyes situated on the surface of the prostomium. Each eye constis of a cup-shaped structure with a pigmented wall made up of radially arranged long, narrow *retinal cells.* Each retinal cell is distinguished into three parts:

(a) An outer nucleated part drawn out into a nerve fibre of the optic nerve.

(b) A highly pigmented middle part or the main body.

(c) An inner part forming a transparent cuticular rod of the hyaline layer.

The cup is filled with a gelatinous matter forming the *lens*, which, according to *Andrews*, is secreted by the retinal cells. The outer exposed general surface of the eyes is covered with epidermis, whose cells are more or less flattened, forming *cornea.* The cuticle of the general surface also covers the eye. Retinal cells are continuous with the epidermis at the edges, thus, forming a small opening of the cup towards cornea, which functions as a *pupil.* The retinal cells join by means of processes with the brain.

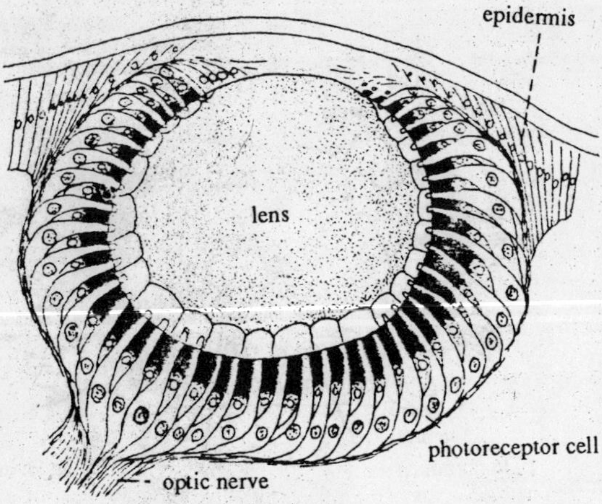

Fig 2.17 *Neanthes. V.S. of eye*

Reproductive System.

Neanthes is unisexual and the male and female may exhibit slight sexual dimorphism. The gonads i.e. ovaries and testes are temporary organs, appearing only during breeding season. The breading season is summer. these gonads disappear after the liberation of gametes. In male *N. dumerilli* there is only one pair of testes lying in each segment between the 19th and 25th segments in one of the segments. But in *N. virens* the testes extend in many sagments around blood vessels. The gonoducts are absent.

During breeding season the cells which detach from the coelomic epithelium in male constituting the spermetogonium becomes disseminated throughout the coelom. These cells undergo rapid divisions to form spermatids which undergo spermiogenesis to form spermatozoa. A mature spermatozoon consists of a minute rod-shaped *head* and a vibretile *tail*. Similarly the oogonia in female separate into the coelomic fluid and after divisions form the ova which are found floating in it.

The gametes are esceped through the nephridia or by the rupture of body wall. A pair of dorsal *ciliated organs* are found in each segment of the body. Each ciliated organ ia a small, ciliated tract of folded funnel-shaped structuer which opens into coelom by means of a wide aperture. It is believed that they are like coelomoducts and act as gonoducts. During breeding season, they open ot the outside by temporary apertures.

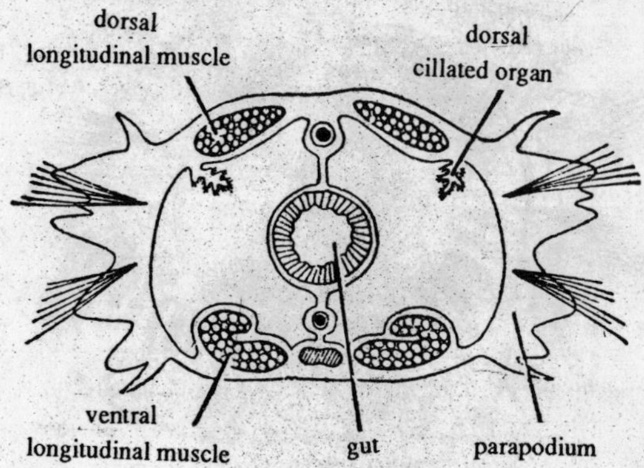

Fig 2.18 *Neanthes*. T.S. of a segment showing ciliated organs

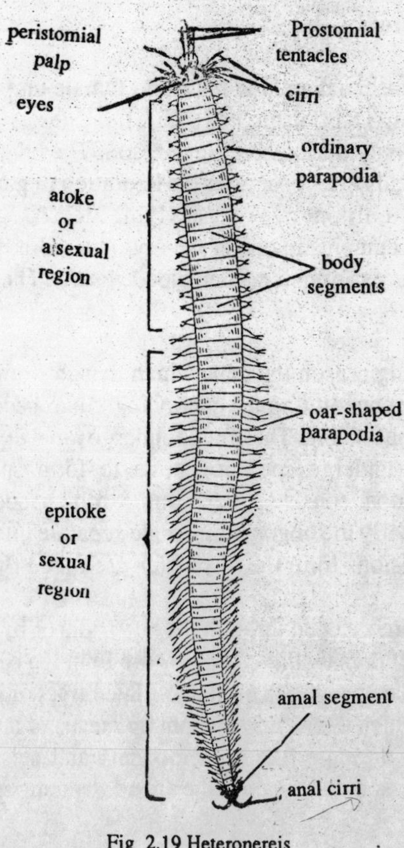

Fig 2.19 Heteronereis

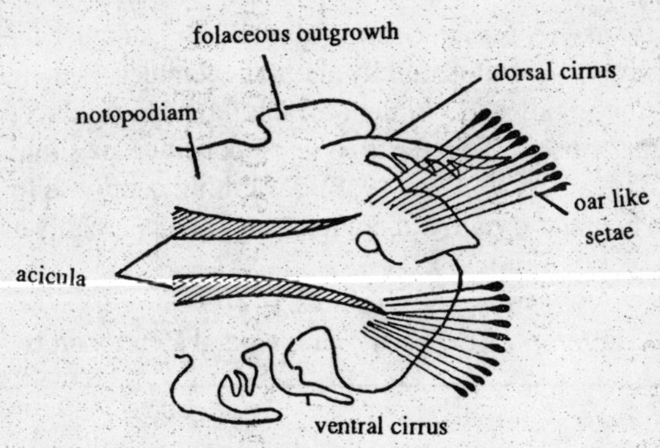

Fig 2.20 Parapodium of Heteruereis.

Heteronereis phase.

N. dumerilli, undergo remarkable changes in external form and behaviour during breading season. At sexual maturity, most of the posterior segments are filled with gametes. The parapodia are also modified. The notopodia, neuropodia and cirri are enlarged and the formation of additional foliaceous outgrowths occur. The old setae are replaced with much longer, more numerous and our-shaped setae which are arranged in a fan-like manner. The old parapodial muscles are also replaced by new muscles. This posterior region of body is sexual region or *epitoke* where as the anterior non-sexual region is referred to as *atoke*.

Following characteristics are noted in heteronereis phase :

1. Heteronereis does not crawl on sea bottom but it swims actively on water surface.

2. The eyes become enlarged and the animal properly precieves the light stimulus.

3. The peristomial cirri are comparatively larger than *Neanthes*.

4. The intestine becomes compressed and functionless due to much development of the gonads.

5. Special sensory papillae are formed on the anal segment.

Swarming

Neanthes swarms on moonless night during summer months after sunset. The males swarms rapidly about the females. Females peoduce some substance called *Ferlilium* (not the fertilizin produced by eggs) which attracts males. This fertilizin is responsible for stimulating and shadding of gametes (*Barnes*, 1968). Swarming continues for about an hour or so. Females die after liberation of eggs and probably makes also.

Table 2.1
Defferences between *Neanthes* and Heteronereis

Neanthes	*Heteronereis*
1. *Neanthes* is a non-sexual sluggish creature, creeping at the bottom of sea.	1. Heteronere*is* is an active swimmer on the surface of water.
2. Eyes normal and the worm is less sensitive to light.	2. Eyes become enlarged and conspecuous so that the heteronereis is very sensitive to light.
3. The parapodia are thin, flap-like and non-vascular.	3. The parapodia of the posterior or the sexual part are large, highly vascular and develop leaf-like outgrowths.
4. The setae on the parapodia are normal.	4. The normal setae are replaced by large, flattened and oar-shaped setae.
5. Setae are arranged in the form of bundles.	5. Setae are arranged in fan-like manner.
6. Sensory projections are well developed.	6. Sensory projections become shruken and the anal segment develops special sensory papilae.
7. Dorsal cirri are less curved.	7. Dorsal cirri become altered.
8. The intestine is well formed.	8. The intestine is compressed and becomes functionless due to development of gonads.
9. Less developed for swimming.	9. Better adopted for swimming.

Fertilization.

Fertilization is external and takes place in sea water except *Platynereis megalops* where internal fertilization takes place. Here the male inserts its and segement in the mouth of female. The sperms are passed into the coelom as the intentive is eroded. In the coelomic fluid fertilization takes place.

Development

The development starts after fertilization of eggs. Each egg is covered in a thick transparent gelatinous envelope. Inside this membrane there are two membranes :

(i) Outer membrane is very thin and delicate.

(ii) Inner membrane is thicker and distinctly striated radially, therefore, it is known as *zone radiata.* The cytoplasm contains numerous yolk spherules and oil drops.

Fertilization produces several changes in the egg. The yolk spherules move at the lower side or vegetal pole, leaving a granular protoplasmic area at the upper or animal pole.

Cleavage. The first two cleavages of the zygote are the equal and vertical and result in four cells or *blastomeres* lying in the same plane. The cleavage is *determinate,* i.e., the fate of the balstomere is fixed and after the four-cell stage, each blastomere gives rise to only one quadrant (quarter) of the embryo. The third cleavage is unequal and horizontal (i.e., at right angles to the first two cleavages) and produces four small *micromeres* towards the animal pole and four large *macromeres* towards the vegetal pole. The remaining cleavages are also unequal. The fourth, fifth and sixth cleavages are also horizontal and cut off three more quartettes of micromeres from the megameres. The cells of adjacent quartettes lie alternately. This type of cleavage is termed *spiral cleavage.* One micromere of the second quartette and one of the fourth are larger than the others. These are respectively termed the *first* and *second somatoblasts.*

Fore-runners of germinal layers are ready now. The micromeres will give rise to the ectoderm, the macromeres to the endoderm and the

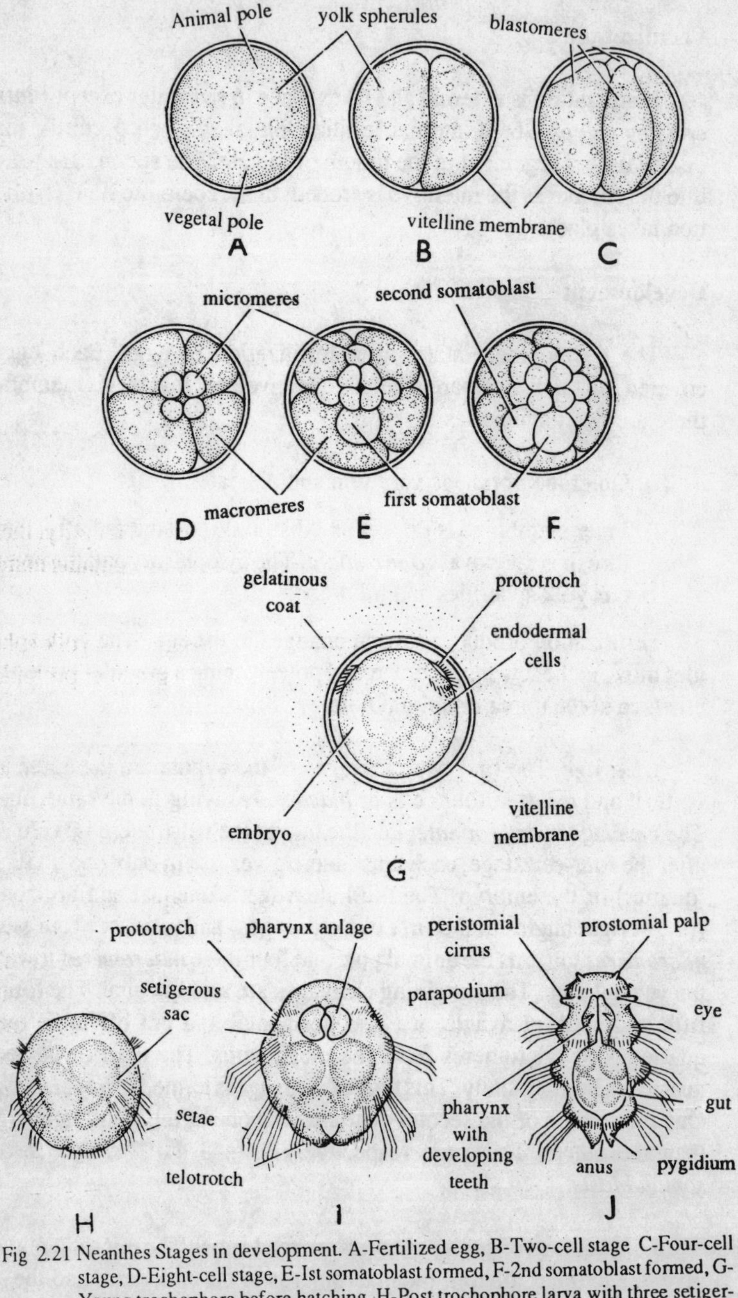

Fig 2.21 Neanthes Stages in development. A-Fertilized egg, B-Two-cell stage C-Four-cell stage, D-Eight-cell stage, E-Ist somatoblast formed, F-2nd somatoblast formed, G-Young trochophore before hatching, H-Post trochophore larva with three setigerous segments, I-Older larva, J-Later larva, after three weeks

second somatoblasts to the entire mesoderm and a small part of endoderm.

Gastrulation. Micromeres divide and redivide to cover four micromeres and descendants of somatoblast, except at blastopore. On the closing of the blastopore gastrulation is completed.

Development of Trochophore larve. At the top of presumptive end, the first quartest of micromeres differentiates into *apical plate,* which differentiates into *cerebral ganglia* and a tuft of *cilia.* A tuft of cilia develops slightly behind the apical plate and is known as *prototroch.* Stomodaeum (rediment of mouth and oesophagus) develops as an ectodermal invagination and a similar invagination at the position of blastopore forms the *proctodaeum.*

The cells drived by the multiplication of first and second somatoblasts are arranged on the presumptive ventral side of the embryo and form the *ventral plate.* The ectodermal cells present along the ventral side are thickened into *ventral nerve cord.* The deeply situated derivatives of somatoblast give rise to *mesodermal bands* which develop into muscles of body wall and of alimentary canal.

The development starts while the egg floats on water surface.

Trochophore larva. The trochophore is rounded and transparent, it has a thin external ectodermal epithelium which is thickened at the two ends and along an equatorial ring. There is a curved *gut* with a mouth, ectodermal *oesophagus* or stomodaeum, an endodermal *stomach,* and an ectodermal *hind gut* opening by an anus. On the thickened parts of the ectoderm is an anterior ciliated *apical organ* with an *apical ganglion* below which is an *eye spot,* at the posterior end are some large cilia and on the equatorial ring is a pre-oral ciliated band or *prototroch.* Between the ectoderm and gut is a large cavity, the *blastocoel* having *mesenchyme cells* larval *mesoderm* and a pair of larval *nephridia,* each made of two hollow cells, one of which contains a flame of cilia. There lies an otocyst near the nephridia. The trochophore is a small, ciliated larva which is without segmentation and also lacks coelom.

An important point to be noted in the development of Neanthes is the gact that the fate of every cell in the early embryo is predetermined. Study of following the fate of cells is called *cell-lineage.*

In *Neanthes,* the trochophore stage is passed within the egg membrane and the ciliated *nectochaetes larva* comes out of the egg. It possesses three segments and bristles.

Metamorphosis. The nectochaetes larva swims about freely for a few days, feeding on microorganisms and then undergoes metamorphosis to become a young worm. Metamorphosis begins with the elongation of the lower region becomes constricted into segments which soon develop bristles. The ciliated bands disappear and the upper part of the larva becomes the head. The part of larva behind prototroch elongates and develops on its sides two pair of invaginations called *setigerous sacs.* Long setae grow out from the setigerous sacs. A third pair of setigerous sac with projecting setae is also added. The mesenchymal cells multiply and form a pir of mesodermal bands one on either side of the gut. Superficial division of body into segments is accompained by internal segmentation of mesodermal bands into three pairs of somites. Each somite develops a cavity, the *coelom.* Both the somites of each segment unite above and below the gut forming the dorsal and ventral mesenteries. Later on the ventral mesentery is disappeared. The coelom obliterates the blastocoel. The part anterior to prototroch modified into the prostomium. The part behind prototroch gives rise to the peristomium. The part bearing the anus is continually pushed backwards as new segments are formed in front of it. Tentacles, palps, parapodia with their cirri and permanent setae are formed. Larval setae are dropped. Larval nephridia are replaced by permanent nephridia. The young worm settles down on the bottom and changes into the burrowing abult worm.

Significance of Trochophore

The trochophore larva has much significance since it is found widely in different groups such as Turbellaria, Mollusca, Bryozoa, Nemertines, annelida etc. It has much resemblence with adult rotifers. The early development of the animals of these groups is almost identical and their trochophore stages are also similar in many respects. The differences in their development arise after the trochophore stage. The affinities of annelid trochophore can be discussed as follows:

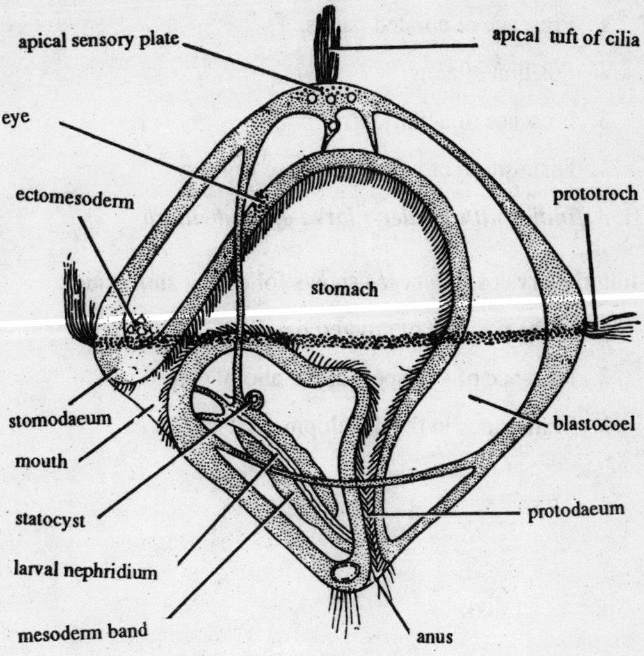

Fig 2.22 *Neanthes* A typical trochophore larva.

I. Affinities with adult ctenophore similarties

1. Both have oval or pear-shaped body.
2. The apical sensory plate is compared with the aboral sense organ of ctenophore.
3. Subectodermal radiating nerves are comparable in both the cases.

Differences

1. Anus is present in trochophore but absent in ctenophores.
2. Cleavage pattern is different in both.

II. *Affinities with Rotifera*

The adult rotifer shows following features of similarities:

1. Presence of ciliated bands.
2. Position of anus
3. Presence of nephridia.
4. Disposition of brain and sense organs.

III. ***Affinities with Muller's larva of Trubellaira***

Muller's larva of *Planocera* shows following similarities:

1. In the position of ciliated bands.
2. Presence of eye-spots at the aboral end.
3. Similarities in the development.

3

PHERETIMA POSTHUMA

The earthworm belong to the class Olgochaeta and are represented by about 1000 species. These are cosmopolitan in distribution except Artic and Antarctic regions. There are several genera of earthworms including *Lumbricus* of Europe and North America, The common Indian species is *Pheretima posthuma.* It is found in South-East Asia, Japan SriLanka and Australaia. *Drawida* is the chief genus of South India. *Megascolex* found in South, India, Sri Lanka, Australia, North New Zealand and Tasmania. *Eutyphaeus* is restricted to Gangatic plain. The genus *Pheretima* has nearly 500 species of which 13 are exclusively Indian. *Prfessor K.N. Bahl* brought out an elaborate monograph on *P. posthuma* which broadly applies to most species The following description is based on the monograph.

SYSTEMATIC POSITION

Phylum	-	Annelida
Class	-	Oligochaeta
Order	-	Opisthoproa
Genus	-	*Pheretima*
Species	-	*posthuma*

Habitat.

The earthworms live in the moist soil rich in dead organic meters or humus, in the temperate and tropical regions. They prefer loamy or pasthy sandy soil of old pastures, lawns and garden. They do not prefer very clayey or acidic soils. According to *Henson,* one acre of land may contain 50,000 earthworms, on an average. The earth worms prefer moist soil as their habitat for various purposes. Respiration takes place through the skin, which the wet soil keeps moist for the diffusion of gases. In the soil they obtain their food materials. The moisture and humus make the earth soft for burrowing.

Habits

1. *Burrowing.* They lives in burrows for the protection against enemies and unfavourable conditions. It makes its burrow by simply pusingh the body through the soil, if it is soft, or by eating the soil, if it is hard.' The burrow runs almost vertically into the earth and may be upto 45cm. deep and it wall are plastered by the secretion of skin-glands. The deeper end of the burrow is rounded and specious so that the worm can take a turn. In cold weather the opening of burrow is closed by dried leaves and soil. The opening of burrow are either protected by dry leaves or characteristic casting of the worm. During dry season the worm descends much deeper (180-210 cm.) in search of moisture. During rainy season, when the burrow gets flooded with water, it comes out on the surface.

2. *Nocturnal animal.* It is *nocturnal* in habit, lives in the burrow during the day but comes out in the night for food, which consists of decayed leaves, organic and other animal matter.

3. *Feeding.* Earthworm feeds on dead organic matter present in soil. It takes its food along with the the soil by sucking action of its pharynx. Some of the matter is digest in alimentary canal, the undigested residue is eleminated as small pills or pellets on the surface of ground, forming the *worm casting.* The earthworms also feed directly on dead leaves, green algae and other matters.

4. *Breding.* Although earthworms are bisexual but there is no self - fertilisation. Copulation takes place between two individuals and involves mutual exchanges of sperms The breeding season last from July to October. Copulation occurs at night or early in the morning before sunrise. Each worm lays its ova and sperms in an egg - case or *cocoons* are laid in moist soil. A single worm develops in one cocoon.

5. *Regeneration.* There is considerable power of regeneration in earthworms by which they produce some segements which are removed or damaged.

6. *Grafting.* It is a process by which a small piece of animal is inserted into another worm or its any other part so as to form organic union. A very small worm can be formed by cutting away from its middle a piece of several contimeters and grafting the two ends together.

7. *Enemies.* There are several natural enemies of earthworms. These are frogs, birds, hedgehogs, centipede etc. Bacteria and *Monocystis* are also harmful to these animals.

External features

Shape, size and clour. Pheretima posthuma has a long narrow body which is bilaterally symmetrical. The anterior end is tapering while the posterior end is more or less blunt. The thickest part of the body is always situated a little behind the anterior. The dorsal surface of the body is easily distinguished by the presence of a dark median linc of dorsal blood vessel which runs throughout the length of the body just below the skin, while the ventral surface is distinguished by the presence of genital openings and papillae in the anterior part of the body. Size varies from species to species and individual to individual of same species. A mature earthworm, *Pheretima posthuma* may attain the size upto 150 mm in length and 3 to 5 mm in thickness.

The colour of most species is dark brown, the dorsal surface is bit darker then the ventral.

Segmentation. The body is divided into 100-120 segments called *metameres.* Externally the metameres are marked by circular grooves called *annuli.* Similarly, internally the segments are marked by partitions, called *septa.*

Peristomium. A distinct head is absent in the earthworm. The first segment is called *peristomium* containing mouth. The peristomium is prolonged anteriorly into a small fleshy lobe, the *prostomium.* The prostomium is of *proepibolus* type as it slightly encroaches upon the peristomium, between both i.e. persitomium and prostomium there is a distinct groove. The eyes, tentacles as present in *Neanthes* are absent here.

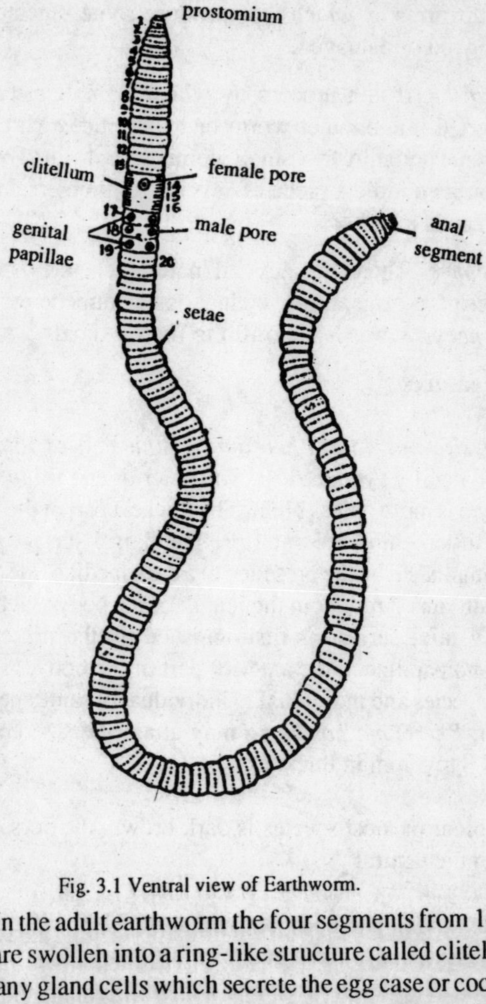

Fig. 3.1 Ventral view of Earthworm.

Clitellum. In the adult earthworm the four segments from 14th to the 17th segments are swollen into a ring-like structure called clitellum. The clitellum has many gland cells which secrete the egg case or cocoon. The clitellum divides the body into three regions, namely anterior *pre-clitellar region,* the middle *clitellar region* and the posterior *post-clitellar region.*

Setae: In all segments except the first, last and clitellum there is a ring of chitinous setae lying embedded in the middle of each segments, setae project beckwards. Each seta arises from a setigerous sac of the skin and is a pale-yellow, curved S-shaped pointed rod-like with a swollon middle part called *nodulus*. In *Pheretima* the setae are arranged in a ring in each segment. Such arrangement is known as perchaetine arrangement. The setae help in loocomotion by holding the earth since they are directed backwards. The movement of the setae is controlled by special type of muscles. With the help of these special types of muscles the setae can be moved in any direction and extended or withdrawn at the will of the earthworm.

Anal segment. The last segment of the body bearing the slit- like opening, the *anus*, *is*

Apertures. The earthworm has following apertures:

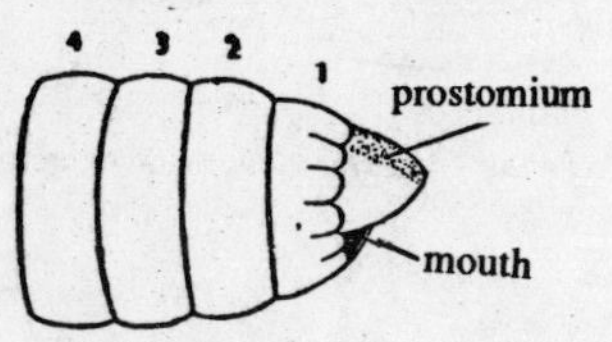

Fig. 3.2. *Pheretima*. Anterior end

(i) *Mouth*. The crescentic mouth is located at the anterior end of the peristomium. This segment shows longitudinal wrinkles and its dorsal edge projects forwards above the mouth as a fleshy lobe called prostomium.

(ii) *Anus*. It is a slit-like aperture located at the posterior end of the anal segments.

(iii) *Female genital pores*. A single median female genital pose lying on ventral side of 14 segment.

(iv) *Male genital apertures*. A pair of male genital pores lies on the 18th segment. In line with the male genital pores, the 17th and 19th sements bear a pair of *genital* or *copulatory papillae*. Each papilla has a shallow pit which looks like an aperture.

(v) *Spermathecal pores*. These are four pairs of sper mathecal openings located on the ventral side in the grooves between the segments 5/6, 6/7, 7/8, 8/9. Through these pores spermatozoa are received from other worm during copulation.

(vi) *Nephridiopores*. These are minute pores which are scattered irregularly over all the segments, except the first two. Through these pores excretory organs called *nephridia* open out.

(vii) *Dors pores.* These are miunte openings lyings on the mid-dorsal line in the grooves between the segments. The first dorsal pore is located between the 10th and the 11th segments. The dorsal are the openings of the coelom to the exterior.

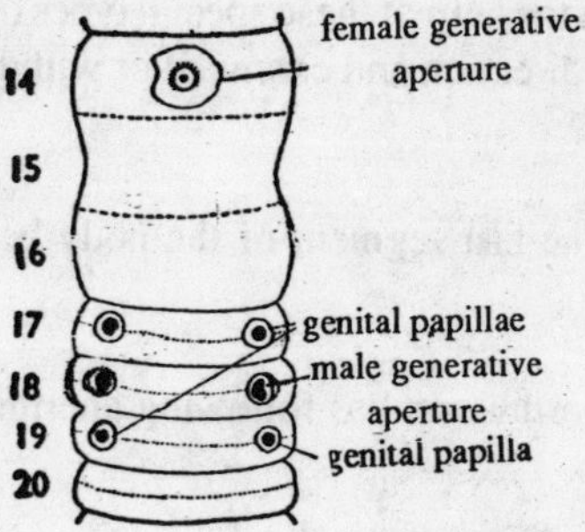

Fig. 3.3 *Phereting* - Ventral view showing genital area

Body wall

The skin or body-wall is thin, soft, elastic, delicate and moist and consist of following layers from outside:

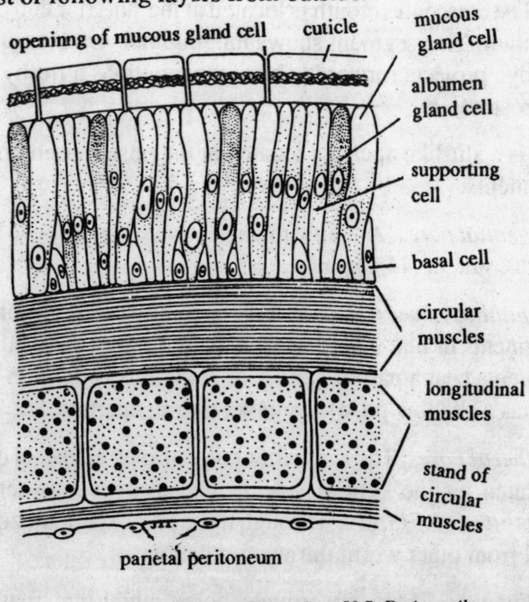

Fig. 3.4. *Pheretima.* V.S. Body wall

Cuticle. The cuticle is thin membrance investing the whole body. It is entirely non-cellular in nature and consists of an inner and an outer transparent layer and a thinner and deeply staining midle layer. It bears many performations through which the epidermal mucous gland cells open out. It is made up of a collagenous protein a polysaccharide along with a little amount of gelatin. The cuticle is secreted by the supporting cells of the underlying epidermis.

Epidermis. The epidermis is a single layer of cells lying beneath the cuticle. It consists of four types of cells, *gland cells, supporting cells, basal* cells and *sensory cells.*

(a) *Gland cells.* The gland cells are of two types:

(i) mucous cells and (ii) albumen cells.

(i) *Mucous cells,* which are numerous and ovoid. Each mucous cell has a broad distal end and a narrow proximal end, containing cytoplasm and nucleus.

(ii) *Albumen cells.* There are few in number and each has secretory granules distributed throughout the cells and the nucleus in the proximal end.

(b) *Supporting cells.* The supporting cells are tall and narrow with oval nucleus in the middle of each cells. They form the bulk of epidermis.

(c) *Basal cells.* These are small, conical and rounded with a nucleus. These cells are situated between the inner ends of supporting and gland - cells.

(d) *Sensory or Receptor cells.* These cells are found in groups and receive the stimuli, hence are known as receptor cells. In appearence they are like the supporting cells but bear hair- like process at their outer ends.

Musculature. It lies beneath the epidermis and consists of an outer thin layer of *circular muscle fibres* and an inner thick layer of *longitudinal muscle fibres.* The longitudinal muscles fibres are arranged into discrete bundles separted by connective tissue septa. These bundles runs longitudinally. In the zones of setae the bundles of longitudinal muscles are separated by setae.

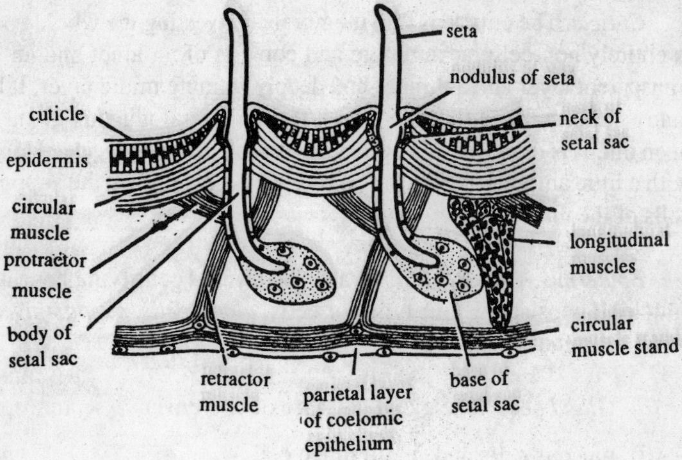

Fig. 3.5. *Pheretima*. A Part of body wall showing setae

The muslce fibres are unstriped, long and spindle-shaped. Pigment cells, connective tissue, nerve fibres and blood copillaries are found amoing the circular muscle fibres.

Musculature of setal sac. The setal sac arises as an invagination in the epidermis and is distingushed into a base, body and neck. To the basal part of each setal sac are applied two sets of muscles, the *protractors* and *retractors*. The protractor muscles, pasing outwards to join the circular muscles, passing outwards to join the circular muscles while the retractor muslce passing inwards to join a thin sheet of circular muscles which lies immediately above the parietal layer of coelomic epithelium. These muscles are responsible for the movement of setae during locomotion. The setae remain and secreted by the large cell of setal saces.

Functions. The body wall performs several funtions which are:

1. It protects the internal more delicate structures from mechanical injury.
2. The cuticle checks excessive evaporation.
3. It is thin, moist and vascular, through which exchange of gases takes place.
4. Mucous secreted by mucous glands of epidermis keeps the body surface clean, free and slimy and cements the walls of the burrow of the worm.

5. Receptor or sensory cells receives the external stimuli.
6. Satae, the organs of locomotion, are lodged in the body wall.
7. The albumen secreted by albumen glands serves as the food for the worms developing inside the ootheca.

Coelom

It is a large perivisceral cavity between the body wall and alimentary canal. It is lined by *parietal epithelium* on the outside and by *visceral epithelium* on the inner side. Each epithelium consists of a single layer of flat or pavement cells. The coelom is not a continuous cavity, but is partitioned into numerous chambers by a succession of transversely placed septa, each of which is inserted into the body wall on the one hand and into the wall of alimentary canal on the other. Each septum consists of a thin layers of interlacking muscle fibres sandwitched between two layer of coelomic epithelium. Certain striking features about the sep may be noted. There are no septa dividing the coelom of the first four segments, the first spetum of the series being observed between the segments 4 and 5. A few muscular strands, however, pass from the wass of the buccal cavity and the pharynx to the body wall. There is no septum between segment 9 and 10. Septa 5/6, 6/7. 7/8, 8/9 and 10/11 are exceptionally thick, muscular and cone-like and run obliquely backwards from the body wall to gut wall. The first few septa (upto the septum 13/14) are complete and unperforated, but from the septum 14/15 backwards all the septa are perforated by minute apertures. The sphincter muscles regulate the opening of these apertures.

It is thought that the coelomic fluid can be restricted to certain segments by the closure of these apertures, which make those segments turgid and adds to the firmness with which the setae are fixed to the ground during locomotion.

Coelomic fluid. The perivisceral cavity is filled with a colourless or milky alkaline fluid known as *coelomic fluid*. The fluid consists of four types of coelomic corpuscles which are as follows:

(i) *Circular nucleated cells*. These are round and nucleated and possess clear protoplasm. There are also some markings on the general surface and these are about ten percent of the coelomic-corpuscles.

(ii) *Amoeboid corpuscles* or *phagocytes*. These are numerous and the largest. Each has a deep concavity on one side and is provided with several folds. Each phagocyte has also many ingested granules e.g., bacteria.

(iii) *Chloragogen or Yellow cells*. In number, they are as many as phagocytes but are smaller in size. Each cell gives out numerous vesicular projections.

(iv) *Mucocytes*. These are elongated cells, each consisting of a narrow nucleated body with a broad fan-like structure.

The major part of plasma, in which coelomic corpuscles are found consists of water, some proteins and salts.

Coelom opens to the outside through dorsal pores and ejects coelomic fluid which kills myriads of bacteria of the soil.

Functions. The coelomic fluid performs a number of funtions which are:

(i) The coelomic fluid helps in distributions of digested food material.

(ii) It keeps the body moist for respiration.

(iii) It destroys bacteria and harmful parasites.

(iv) It aids in excretion by eleminating yellow cells, which are full of waste products.

(v) The luminosity produced by some earthworms is due to the coelomic fluid.

(vi) It helps in locomotion by causing turgidity.

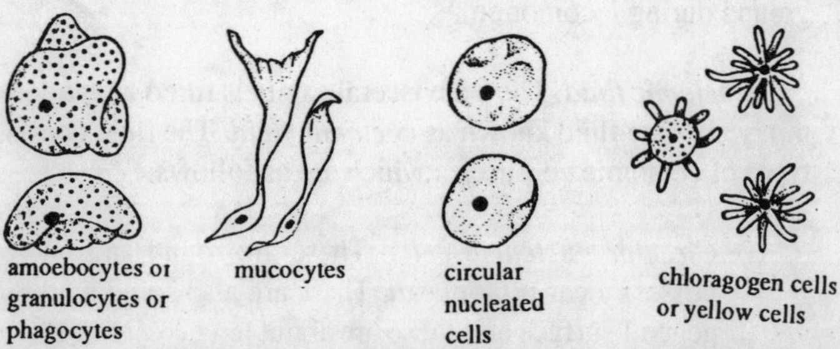

Fig. 3.6. *Pheretima*. Coelomic corpuseles

Locomotion

The earthworm moves by the contractions of the musculature of body wall aided by setae. When an earthworm starts to crawl, the first few segements become thinner but longer. This is due to contraction of circular muscle in that region, elongation being directly caused by an influx of coelomic fluid, which is driven away from the contracting regions into more anteriorly placed segments. The setae of these segments protrude and get a grip of the substratum. The longitudinal muscles of the anterior region now contract, making the region short and thick. As the anterior region is frmly anchored, its shortening pulls the hinder region forwards. These in turn pull the next few segments and so on. These alternating series of waves of extension and contraction aided by the leverage afforded by the setae bring about the locomotion of the worm. Coordination of the waves of muscular contraction is maintained by reflexs passed on from one segment to next by mechanical stimulation and by inpulses passing along the ventral nerve cord. The contraction of both the muscles are also controlled by reflexes.

In order to illustrate the sequence of locomotion in earthworms, *Gray* and *Lissman* (1938) figured a hypotheical earthworm consisting of only 20 segments. It those segments in which the longitudinal muscles are contracted, the segment as a whole is at its thickest and shortest with setae protruded and in contact with the ground. The segments in which the circular muscles are contracted are at their most elongate and thinnest, have their setae withdrawn and are not in contract with the ground.

The body of earthworm moves forward more rapidly. The worm covers a distance of about 25 cm. in one minute. However, it can also move backwards by reversing the direction of the setae. When disturbed, the earthworm performs violent, irregular and jerky movements. The worm can not move ovar a smooth hard surface like glass plate.

Digestive System

The digestive system comprises:

I. Alimentary Canal

II. Digestive glands

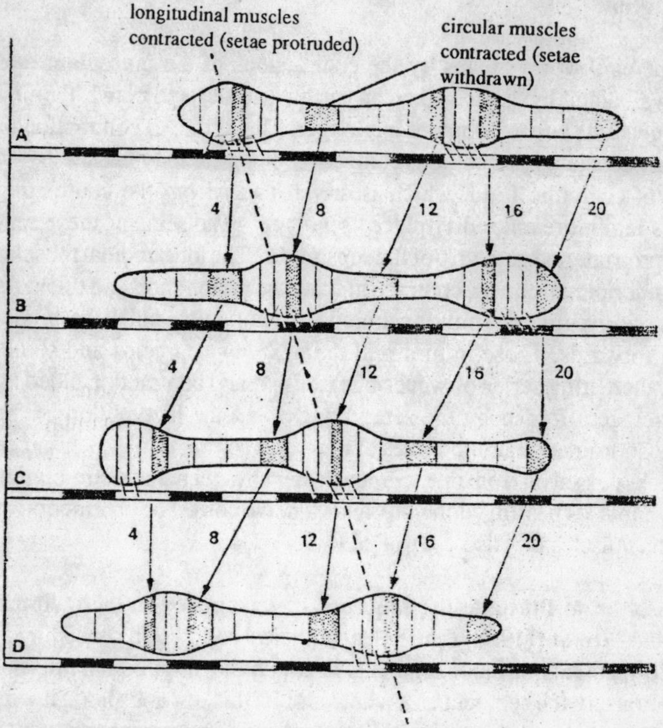

Fig. 3.7. *Pheretima* Showing locomotion.

I. *Alimentary canal.* The alimentary canal is complete. It is straight tube of varying diameter. It runs through the entire length of the body from the mouth at the anterior end to the anus at the posterior end. It is held in position by the successive transverse intersegmental septa. The alimentary canal comprises following parts:

1. *Mouth.* The mouth is a crescentic aperture situated on the anterior face of first sement, the peristomium. It is over hung by a small fleshy lobe, the prostomium. It opens into the buccal chember.
2. *Buccal chamber.* The buccal chamber extends up to the middle of the 3rd segments. Its walls are slightly folded. The buccal chamber is protrusible and as such as an organ of ingestion of food. The buccal chamber leads into pear-shaped structure, the pharynx.

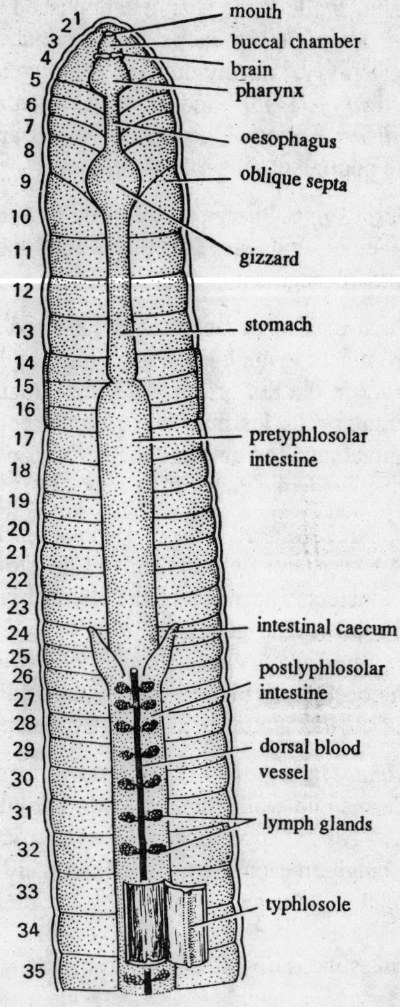

Fig. 3.8 *Pheretima*. Alimentary canal

3. *Pharynx*. The buccal chamber leads into a spacious pear-shaped muscular *pharynx*, which extends up to the 4th segments. Its anterior end is marked by a nerve ring placed in a groove between it and the buccal chamber. Its cavity is somewhat dorso-ventrally compressed due to the presence, on its dorsal roof, of a large glandular *pharyngeal mass* producing a salivary secretion. The

lateral walls of the pharynx are pushed inside forming a narrow *horizontal shelf* on each side. The two shelves meet anteriorly and posteriorly, thus dividing the pharyngeal cavity into a dorsal or *salivary chamber* and a ventral or *conducting chamber.* The *salivary secretion,* containing *mucus* and proteolytic enzymes, are poured into the salivary chamber.

4. *Oesophagus.* Pharynx is followed by a straight narrow, long tube called oesophagus. This thin walled tube extends from the 4th upto 7th segment.

5. *Gizzard.* The oesophagus is followed by the gizzard. It is a prominent, oval, hard, thick-walled and highly muscular organ lying in the 8th or 8th and 9th segments. It has thick wall of circular muscles lines by columnar cells. Inside the columnar epithelium is a lining of cuticle. The food is ground up into a fine state inside the gizzard.

6. *Stomach.* Gizzard leads into stomach, which extends from 9th to 14th segment. Both the ends of the stomach are provided with sphincters. The walls of the stomach are folded, vascular and glandular. The glandular folds have *calciferous glands* found only in *Eutyphoeus* and *Lumbricus.* These cells produce a calcareous fluid, which neutralises the acidic effect of decomposing leaves, that are taken as foods by the earthworm.

Intestine. The region next to the stomach is the *intestine* which is a long, wide and thin-walled tube extending from the 15th segment to the last. It has a beaded appearance due to constricutions corresponding to septa but bulgin in each segment. the internal lining is ciliated, folded, vascular and glandular.

The intestine is divisible into three parts:-

(i) *Pre-typhlosolar region:* The first part of the intestine lying between segments 15 to 26 is the pre-typhlosolar region. In this region the lining of the intestine is folded to form villi which are extemely vascular. In the 26th segment, two short and conical outgrowths, one on either side, are given off from the intestine, these are called the *intestinal caeca.* These extend forwards over three or four segements and have a special blood suply; internally they have many longitudinal folds which appear as villi in transverse sections, epithelial cells being highly vascular and filled with secretory granules, *Chen* and *Puh* (1941) believe that these caeca are digesive glands and secrete an amylatic enzymes which digest starch.

(ii) *Typholosolar region.* This part lies between the 26th segments up to 23-25 segmeents in front of the anus. This is characterized by the presence of a higly glandular and vascular longitudinal ridge, arising as a meian ingrowth of the dorsal aspect of the intestinal cavity. This is called the *typholosole.*

(iii) *Post - typhlosolar region.* It occupies last 23 or 25 segments. It is without a typhlosole and also called as *rectum.* It contains small pallets of mud which are passed out from the anus in the form of castings.

8. *Anus.* it is a slit-like aperture on the posterior face of the anal segment.

Histology

the wall of alimentary canal consists of four layers in succession. These from outside to inside are:

1. *Peritoneum.* The outermost covering is formed by peritoneal epithehium, which consist of tall and narrow cells. In the stomach and intestine some of these cells are full of yellow refractile granules, hece called, *yellow* or *chloragogen cells.*

   Barnes. (1968) mentions that chloragogen cells play a vital role in intermediary metabolism similar to the role of liver in vertebrates. Chloragogen tissue is regarded to be the chief centre of glycogen and fat synthesis. It is also regarded that deamination of protein formation of ammonia and synthesis of urea also occur in these cells. *Semal Van Gansen* (1956) pointed out that in terrestrial species, the silicates got from food material and soil are removed from the body, Histologically, chloragogen tissue is derived from the peritioneum and its colour is due to the presence of green-yellow lipid inclusions. Chloragogen cells are released into the coelom as free cells known as eleocytes. *Liebman* (1946) describes that eleocytes have food reserves and migrate as different tissues of the body including developing egg. *Semal-VanGansen* regarde eleocytes as degenerate chloragogen cells. which are destroyed by phagocytic coelomocytes.

2. *Musculature*. It forms the second layer. It consiss of an outer layer of *longitudinal muscle fibres* and an inner layer of *circular muscle fibres*. Both these layers are well developed in pharynx and oesophagus but ill-developed in intestine. In the gizzard the main bulk of musculature is represented by circular muscle fibres only. All the muscles are unstriped.

3. *Internal epithelium*. It consists of tall columnar cells covered with thin cuticle in buccal cavity. In pharynx the columnar epithelium is ciliated on the roof and devoid of cilia on the floor. Oesophagus has tall columner cells. In gizzard small cells are covered by thick *internal cuticle*. The epithelium in the intestine has long and narrow *absorptive cells* and *glandular cells* have vacuolated appearence

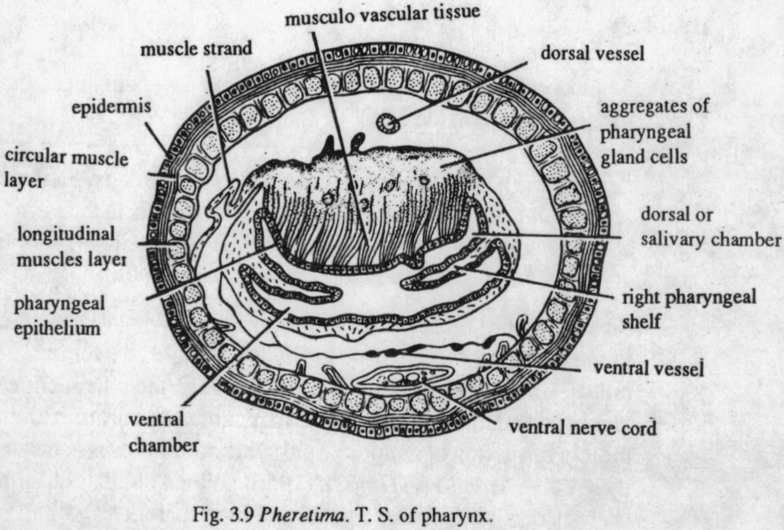

Fig. 3.9 *Pheretima*. T. S. of pharynx.

II. *Digestive glands*. There are three types of digestive glans which are associated with the alimentary canal:

i) *Pharyngeal mass*. It lies over the pharynx and consists of glandule-muscular tissue with abundant bood supply. It secretes a proteolytic enzyme and mucous.

(ii) *Gastric gland*. They occur in the entire epithelial lining of stomach which secrete protelyic enzymes.

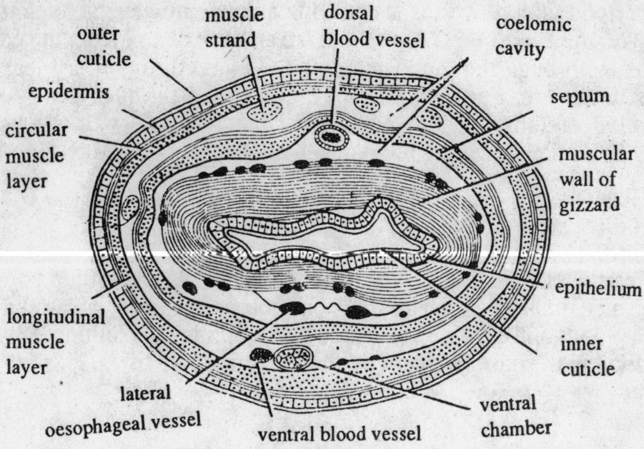

Fig. 3.10 *Phertima* T.S. grazzard

(iii) *Intestinal glands*. They occur in the entire epithelium of inestine. They secrete proteolytic, amylolytic and lipolytic enzymes.

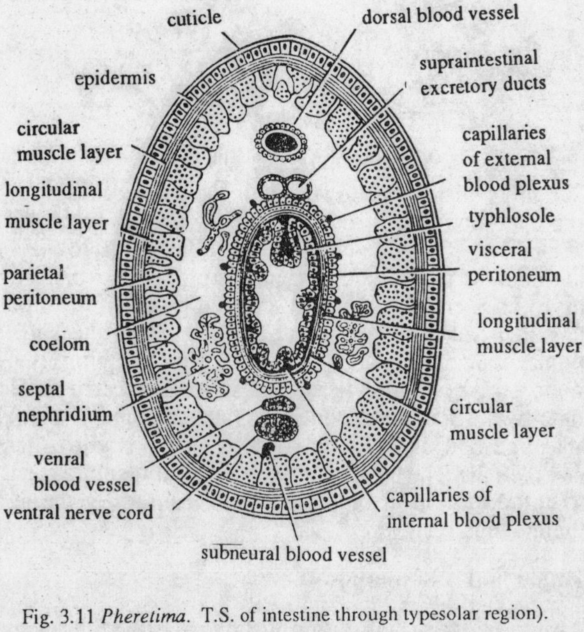

Fig. 3.11 *Pheretima*. T.S. of intestine through typesolar region).

Food and feeding mechanism.

The food of earthworms is decaying animal and vegetable matters which are present in soil. They take large amount of soil from the earth that passes through the gut. They also feed directly upon dry leaves, grasses, algae etc. For taking food, they expand the pharynx by the contraction of muscle-bands connecting the pharynx with the body cavity. The enlargement of pharyngeal cavity draws the food into the buccal cavity by sucking action.

Digestion

In earthworms digestion is extracellulor. The digestive fluid poured in pharynx contains mucin and some proteolytic enzymes. Mucin lubricates the foods and enzyme starts digestion of proteins. Nothing happen to the food in the oesophagus. The food is ground up in the gizzard

A

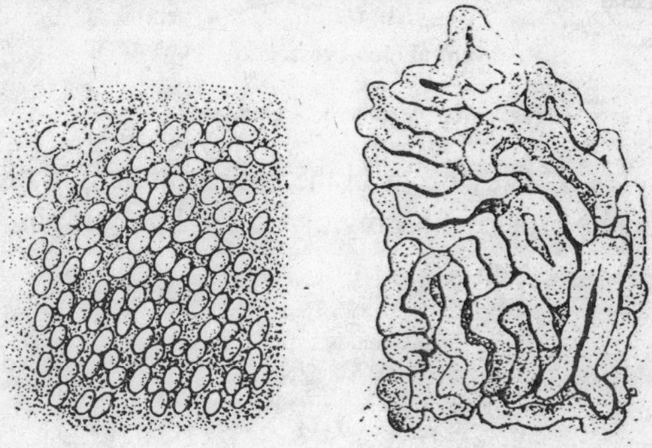

B

Fig. 3.12 Castings A = *Phrtryims* B = *Eutyphoeus*

The soil paticles swallowed with the food probably help in the grinding operation. This grinding may be called *mechanical digestion* since it facilitates the subsequent action of the digestive enzymes. In the stomach another proteolytic enzymes are poured which digest more protein. In *Lumbricus* Calcium is secreted by calciferous gland which probably neuteralizes the humic acid present in the soil. The final digestion of food takes place in intestine. Here proteins are hydrolyzed into amino acids by proteolytic enzymes like trypsin and pepsin, starches into sugar by the diastase and fats into fatty acid and glycerol by lipase. Intestinal digestive juice of earthworm corresponds with pancreatic juice of higher animals. The passage of food in the alimentary canal is caused by *peristaltic movement* of the gut wall, caused by the circular and longitudinal muscles.

Absorption and Assimilation

The digested food is absorbed by the absorptive cells of intestine and is passed ino the blood-stream for distribution. The food is ultimately converted into protoplasm of the cells (assimilation). The typhlosole increases the surface arean for absorption.

Egestion

The undigested food material is passed out with the large quantities of the earth through the anus in the form of casings. The castings of *Pheretima* are in the form of small rounded pallets or balls, while those of *Eutyphoeus* are in the form of heap with an open passage in the middle.

Lymph glands. There are ten glands situated on the dorsal surface of the intestine on both the side on dorsal blood-vessel from 26th segment up to the last segment. Each lobule is a phagocyte or amoebocyte. These glands help in breaking up the waste products including cysts and spores of *Monocystis*.

Respiration

Moist skin is the only organ of respiration. It carries on exchange of gases between the environment and the body. Oxygen diffueses across the outermost cells of the skin. Diffusion continues further across the wall of blood capillaries of the skin to the blood. The blood contains a respirator pigment called *haemoglobin*, which greatly inceases its capacity for oxygen obsorbing. Oxygen combines with haemoglobin to form oxyhaemoglobin. Oxhaemoglobin gives up oxygen once again to tissues which have a low oxygen tension. In the cells oxygen is utlized in the oxidation of food to liberate energy, besides carbon dioxide is also liberated as waste product. Carbon dioxide is diffused out in the blood. Blood with carbon dioxide returns to the skin from where it is diffused out in the surroundings.

Circulatory System

Earthworm has a system of vessels or tubes in which red blood circulates which transports, oxygen, carbon dioxide, digested food and nitrogenous waste to and from the parts of the body. A transportation system, in which blood circulates always in tubes, is called a *closed circulatory system*.

The circulatory system comprises: blood, blood vessesl, hearts and blood glands.

1. *Blood.* The blood is composed of a fluid plasma and colourless corpuscles which can be comparable to the leucocytes of vertebrates. Haemoglobin occurs dissolved in the plasma which gives red colour to the blood.

2. *Blood vessels.* The arrangement of blood vessels is very complicated. The arrangement in the first 13 segment is different from the arrangement of blood vessels in rest of the body. Thus it is easy to describe the circulatory system in the two headings.

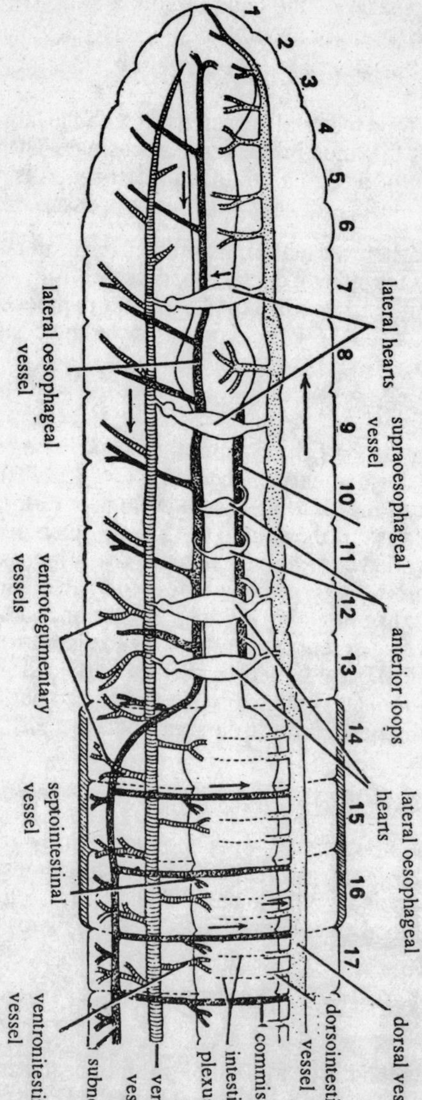

Fig. 3.13. *Pheretima* A diagrammatic representation of the blood vessels in the first 13 segments (anterior region) of the body.

I. Arrangement of the blood vessels behind the thirteenth segment.

II. Arrangement of the blood vessels in the first thirteen segments.

I. Arrangement of the blood vessels behind the thirteenth segments. In this region there are three main longitudinal vessels namely.

(i) Dorsal vessel (ii) Ventral vessel and (iii) Subneural vessel.

(i) *Dorsal vessel.* Dorsal vessel is the largest of the three longitudinal vessels. It runs throughout the length of the body in the mid-dorsal line between the alimentary canal and the body -wall. Its walls are think and muscular. It shows *rhythmical peristaltic contractions*. From behind to the front. The backward flow of blood cannot take place as in this vessel, a pair a forwardly directed valves is found in its lumen in front of each septum. During contraction-movement, the two flaps of valve close down, thus, preventing the backward flow of blood.

In this region dorsal vessel reveives blood and thus serves only a collecting vessel but does not distribute the blood. In each segment, it receives two pairs of *dorso intestinal -vessels* collecting blood from two pairs of *transverse vessles* encircling the intestine. It also receives, in each segment, pair of *commissural* vessels. One on each side, which are ventrally connected with the *subneural* vessel. Commissural vessels run along the poserior face of the septum, one on either side and receive capillaries from the body-wall and nephridia.

(ii) *Ventral vessel.* It is a large vessel that runs mid-ventrally below the alimientary canal and above the nerve cord from the end of the body to another. It walls are thin and non-contractile and the valves are altogether absent. The blood flows through it posteriorly.

The ventral vessel is pricipally a distributing vessel. It supplies blood, in each segment through a pair of *ventro-tegumentary vessels* to the integumentary nephridia, body wall, septa and reproductive organs. Behinds the 13th segment each ventro-tegumentary vessel sends a small branch, a *septo-nephridial branch,* supplying the septal nephridia. Besides these, the ventral vessel, in each segment behind the 13the gives off a median *ventro-intestinal vessel* to the intestine.

(iii) *Subneural Vessel.* The subneural vessal runs mid-ventrally below the nerve cord from the 14th segment backwards. It is a thin vessel. the blood flows backwards in the subneural vessal as in the ventral vessal. It is a collecting vessel and collects blood from the ventral body wall and gives some blood to the intestine.,

II. Arrangement of blood-vessels in the first thirteen segments. In this region the dorsal and the ventral vessels are continued as such up to the anterior end whereas the subneural bifurcates into two lateral oesophageal vessal and a new vessel i.e. supraoesophageal vessel is formed on the oesophagus.

(i) *Dorsal vessel.* In the first 13 segements the drosal vessel acts as distributing vessel. It sends out all the collected blood from posterior region of the body into ventral vessel through the hearts. By a pair of branches in each segments, 3rd, 4th, 5th, 6th and 8th, to pharyngeal nephridia, oesophagus and gizzard respectively. In extends upto the cerebral ganglion in the middle line of the 3rd segments, where it is divide into three small branches supplying blood to pharyngeal mass and the roof of buccal cavity.

(ii) *Ventral vessel.* It extends anteriorly upto the second segment of the body and sends blood to the integument, nephridia, septa and reproductive organs through a pair of *ventrotegumentary vassel* in each segment. Ventrointestinal vessels are absent in this region.

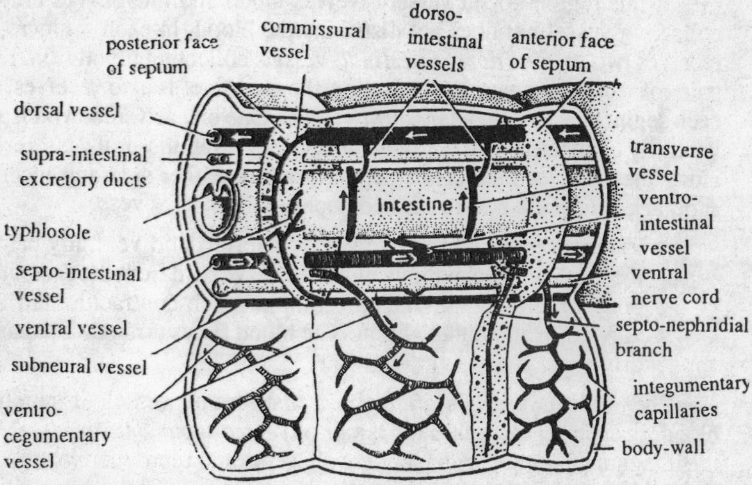

Fig. 3.14 *Pheretima* Blood vascular system in intestinal region.

(iii) Supraoesphageal vessel. It lies on the stomach from the 9th segment to the 13th segements. It is a collecting vessel as it collect bllod from gizzard and stomach. it also receives blood form the lateral oesophageal vessels by two pairs of *anterior loops*. The loops are present in 10th and 11th segments. It sends the collected blood to the ventral vessel through *latero-oesophageal hearts*.

(iv) *Lateral of esophageal vessel.* These are two vesels and lie on either ventro-lateral side of the gut running from the anterior end of the body up to the 13th segment. these receive a pair of *ventro*

- *tegumentary vessels* in each segment, collecting blood from the body wall, septa, nephridia and reproductive organs. Flowing posteriorly, some of its blood passes to the *supra*-oesophageal vessel through a pair of anterior loop in each of the segments 10 and 11 and through several ring vessal running through the wall of the stomach. The rest flows backward into the sub-neural vessel.

(v) *Ring vessels*. The stomach possesses in its muscular wall, some circular vessels of typical nature, known as ring-vessels. These are nearly 12 in number in each segment. They convey the blood from lateral oesophageal to the supra-oesophageal.

3. *Heart*. There are four pairs of hearts, one pair each in 7th, 9th, 12th and 13th segments. They pump blood from the dorsal to the ventral vessel. They possess valves to check backward flow of blood. Since the hearts of 7th and 9th segments connect the dorsal and ventral vessel, they are called *lateral hearts*. They possess four valves each. The hearts of 12th and 13th segments communicate both the dorsal and supra oesophageal vessels with the ventral vessels and are designated as the *lateral oesophageal hearts*. They possess three valves each.

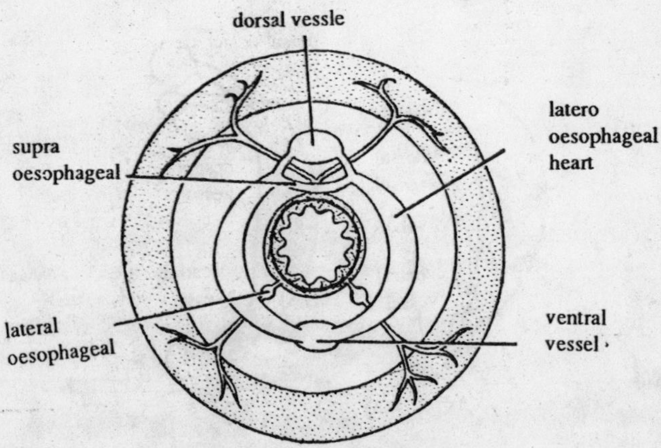

Fig. 3.15 *Pheretima* T.S latteral oesophegeal hearts

4. *Blood glands*. There are several small red follicular bodies just above the pharyngeal mass in each of the 4th, 5th and 6th

segment. These are called blood glands and are connected to the salivary glands. The follicles have a syncytial wall enclosing a capsule containing mass of loose cells. These glands manufacture blood corpuscles and haemoglobin and also supposed to be excretory in nature.

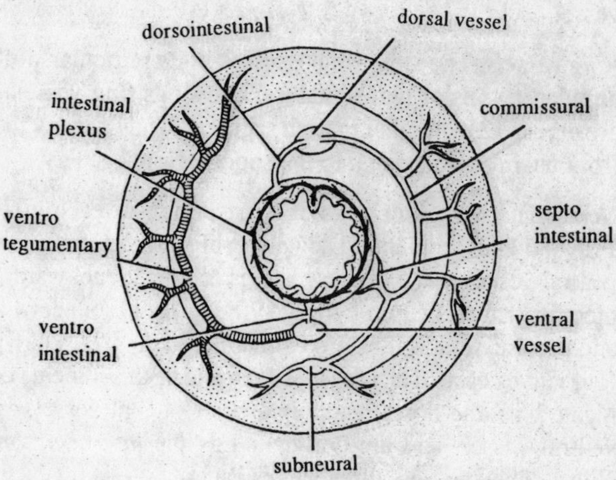

Fig. 3.16 *Pheretima* T.S. though a segment on the left, and through a septum on the light

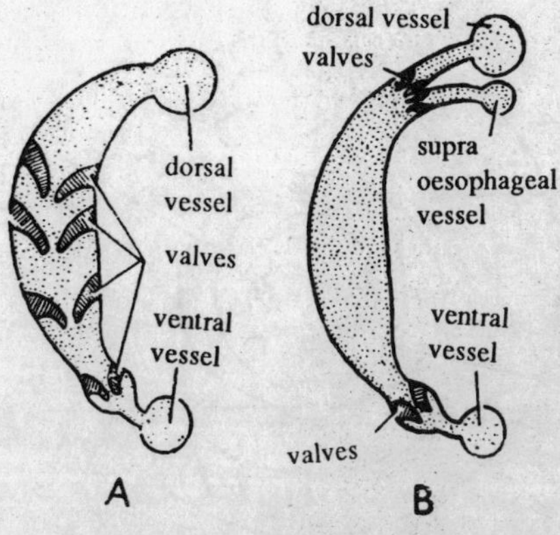

Fig 3.17 *Pheretima*. Diagrammatic representation of hearts in section. A-Lateral heart, B-Latero-oesophageal heart

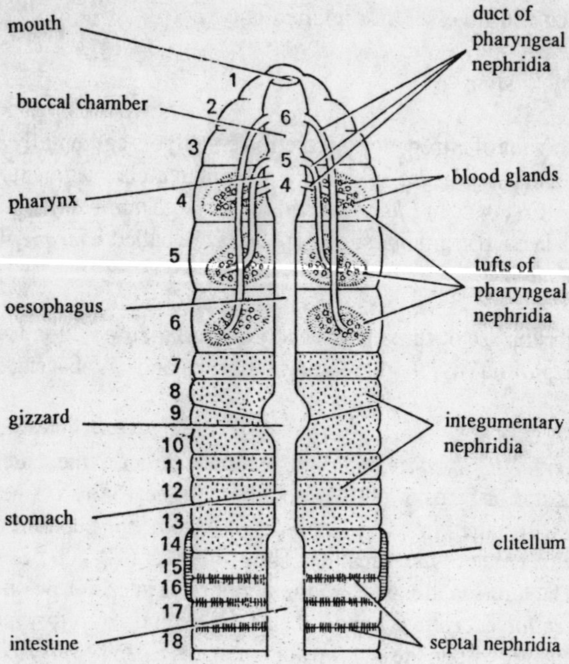

Fig 3.18 *Pheretima*. Different types of nephridia.

Circulation of Blood

The flow of blood is from behind forwards in he dorsal vessel and from in front back wards in the ventral and subneural vessel. The ventral vessel supplies blood to the body wall septa, nephridia, reproductive organs etc. through the ventro-tegumentary vessels and to the gut through ventro-intestinal vessels. This blood is collected by the lateral oesophagea, supra oesophageal and subneural vessels. In the first 13 segments blood from gut, body wall, nephridia, septa, gonads etc. is collected by lateral oesophageal vessels and anterior loops. From gizzard and stomach blood is collected by supra oesophageal which eventually passes its blood to the ventral vessel through lateral oesophageal hearts. In the intestinal region the dorsal vessel collects blood from gut wall through dorso-intestinal and from subneural through the commissural vessels which also receive blood from the body wall, septa, nephridia etc. From the gut wall, the dorso-intestinal vessels carry blood to the dorsal vessel. Thus the dorsal vessel is collecting vessel in the intestinal region where as it is distributing vessel in the first 13 segments supplies blood to the gut and to the ventral vessel through hearts.

Excretory System

The organs of nitrogenous excretion consist of segmentally arranged troubles called *nephridia,* which are present in all the segments of body except the first two. In *Pheretima* there is a large number of small-sized nephridia in each segment, such nephridia are called *micronephridia* or *meronephridia.*

Nephridia are of three types according to their position in body (i) Septial nephridia (ii) integumentary nephridia and (ii) pharyngeal nephridia.

(i) *Septal nephridia* Septal nephridia occur on the anterior and posterior faces of all the septa behind the 15 the segment. They are arranged in two rows on each face of the septum, a row on either side of the intestine. They vary from 80-100 per segment. The terminal ducts of all the nepheridia in each row open into a *septial excretory canal.* The two septial excretory canal run upwards along the septum and open into a pair of *surpra intestinal excretory ducts* which are situated side by idea in the mid-dorsal line above the intestine and below the dorsal vessel extending from 25the to the last body segment. These longitudinal excretory duct open the intestine in each segment through narrow *ductules* near the septum. Since the septial nephridia open into the gut, they are described as *enteronephric.*

Each nephridium consists of three main parts:

(a) Nephrostome or Funnel
(b) Main Body
(c) Terminal Duct

(a) *Nephrostome or funnel.* The ciliated funnel or nephrostome is the structure by which each nephridium opens into the coelom. It is a rounded bilabiate structure, having a slit-like opening by which nephridium communicate with the coelom. The Funnel consists of a central cell and an upper lip of 8 or 9 cells and a lower lip of 4 or 5 cells. It is followed by a short, narrow and ciliated tube, called the neck, which is bet on itself. The *neck* joins the main body of the nephridium.

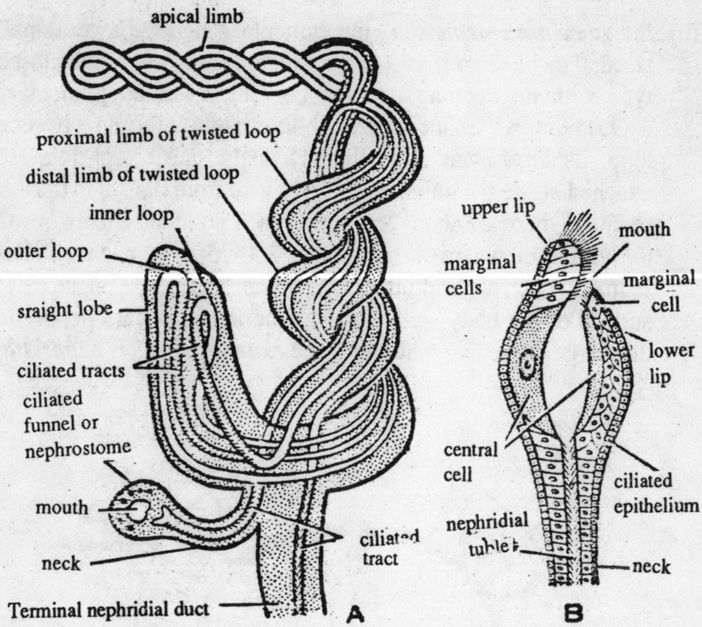

Fig 3.19 *Pheretima.* A-A septal nephridium. B-Detailed structure of nephrostome in L.S.

(b) *Main body.* It consists of a short straight lobe and a long spirally twisted loop. The *straight lobe* is rounded at its distal free end but is continued at its outer end into the twisted loop, which is more than double the length of the straight lobe. The *twisted loop* can be distinguished into proximal and distal limbs, which a re spirally twisted around one another. The distal limb is connected with the straight lobe, while the proximal limb is connected with the neck, ciliated funnel and also given of a terminal nephridial duct. There are four ciliated tracts along the nephridial trouble. The nephridium is made up of a glandular mass. There is a single tube in the neck and terminal duct, while there are four in the straight lobe, three in the basal and two in the apical part of each limb of the twisted loop.

(c) *Terminal Duct.* The terminal ducts of nephridia open into a pair of septial excretory canal, which runs inwards along the septa which are described above.

(ii) *Integumentary nephridia.* Integumentary nephridia are small sized, they have no funnel or nephrostome and are of the closed type with no opening into the coelom. Each integumentary nephridium is V-shaped with a short straight lobe and a twisted loop, its lumen has two ciliated canals. these nephridia are attached inside the lining of the body wall from the 7th to the last segment, there are about 200-500 in each segment, except inside the clitellum where they number 200-2500 in each clitellar segment. Each nephridium opens by a *nephridiopore* on the outer surface of the body wall. Because the integumenary nephridia discharge waste externally they are *exonephric micronephridia.*

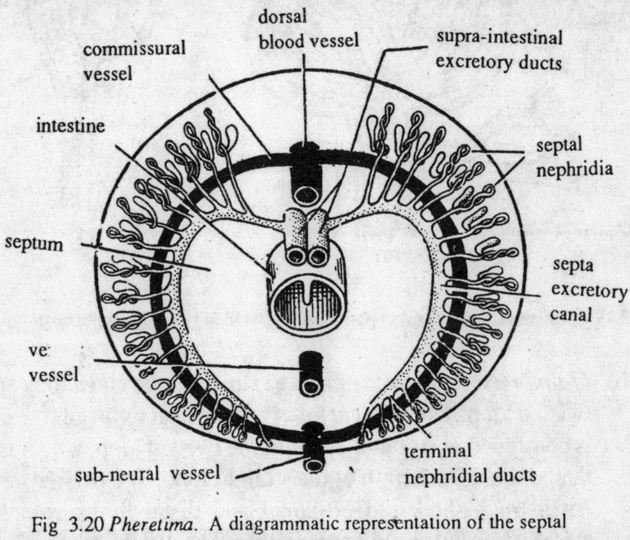

Fig 3.20 *Pheretima.* A diagrammatic representation of the septal nephridial system on a septum.

(iii) *Pharyngeal nephridia.* The pharyngeal nephridia are similar in size and structure to the septial nephridia as described above, but are lacking in coelomic funnels. These nephridia occur in paired massed, in segments 4th, 5th, and 6th lying besides the pharynx and oesophagus. Their terminal ducts join together to give rise ultimately to one pair of ducts in each of the segments 4th, 5th and 6th. The ducts of the pharyngeal nephridia of the segment 6th open in to the buccal cavity in the 2nd segment, whereas the ducts of the nephridia of segments 4th and 5the open into pharynx. The pharyngeal nephridia are interwined with blood glands referred to above.

Physiology of Excretion. Nephridia are chiefly organs of osmoregulation. Nephridia are richly supplied with blood vessels and lined by cilia. The epithelial cells extract waste material, chiefly urea, from the blood. The waste material travel from the body of the nephridium to its terminal duct which discharge it either directly to the exterior or into the alimentary canal.

Working of nephridia covers three functions, filtration, reabsorption and chemical transportation. In open septial nephridia filteration takes place at the funnel, but in other forms of ncphridia i.e. integumentary and pharyngea, filtration takes place through the walls of nephridia. filtration through nephridial wall from surrounding blood vessel also takes place in septial nephridia where its parts are quite narrow. The filtrate consists of blood plasma minus the colloids together with coelomic plasma. As the filtrate passes through the fine vessels in the nephridia large amount of useful substances are reabsorbed like glucose, amino acids, phosphates, chlorides etc. The process is called selective reabsorption. Finally a protein-free urine containing ammonia, urea and allantoin is put out of the nephridia.

The urine of earthworms is *hypotonic* as contains a far lower concentration of salt than do the blood and coelomic fluid.

Some excretory matter got rid off when the coelomic fluid passes out through the dorsal prores.

Nervous System

The nervous system is well developed in earthworms and it comprises of:

1. Central Nervous System.
2. Peripheral Nervous System.
3. Sympathetic Nervous System.

1. *Central Nervous System.* It comprises a pair of supra pharyngeal ganglia, a pair of peripharyngeal connectives, the ventral nerve cord.

A pair of closely united, white, pear-shaped *suprapharyngeal ganglia or brain,* forming the so-called brain, lie dorsally in the third segment in the depression between the buccal cavity and the pahrynx. A paid of thick stout *circum-* or *peri pharyngeal connectives* arise from them laterally, embracing the pharynx and meeting ventrally in a pair of fused *subpharyngeal ganglia* lying beneath the pharynx in the fourth segment. In this way, a complete *nerve ring* is formed around the pharynx.

The *ventral nerve cord,* arising from the sub-pharyngeal ganglia, runs backwards in the mid-ventral line to the posterior end of the body. In each segment, it presents a slight enlargement or ganglion. The ventral nerve cord appears to be single but it is really double, consisting of two compactly united right and left cords as seen in a transverse section. Each *segmental ganglion* also represents the fusion of pair of gangila, one belonging to each cord of the double ventral nerve cord.

The entire nervous system contains nerve cells or neurons but these are more numerious in the gangila. As revealed in a transverse section, the core of the ventral nerve cord consists of two bundles of closely packed nerve fibres, transversely cut in the section. Between these intervene connective tissue fibres. The periphery is thin and contains, in the ventral and later zones, neurons, these are also nerve fibres. In the dorsal zone of the periphery there are four gaint fibres, one medium, one submedian and two lateral. These giant fibres transmit impulses in earthworms.

2. *Peripheral Nervous System.* Various nerves, which are orginated from the central nervous system and supply different parts of the body, comprise peripheral nervous system. Two nerves arise from each side of the brain and innervate the prostomium and the buccal by dividing into a number of branches. Peripharyngeal connectives give off nerves to into a number of branches. Nerves, arising from subpharyngeal ganglia, intervate 3rd and 4th segments. From each ganglionic swelling of the ventral nerve cord arise three pairs of *segmental nerves* out of which one pair is infront of the row of the setae which intervate the gut wall, body wall and other internal organs of their segment. These nerves are of the mixed type consisting of both sensitive or *afferent fibres and* motor *or efferent fibres.* The afferent fibres connect the sensory impulses. Ganglia change the sensory impulse into motor impulse which passes on to the muscle-cells. The sensory im-

pulses may be transmitted to the muscles of the same segment or may travel along the nerve-cord and thus be transmitted to the muscles of other segments. This arrangement serves for the co-ordination of the activities of the different part of the body.

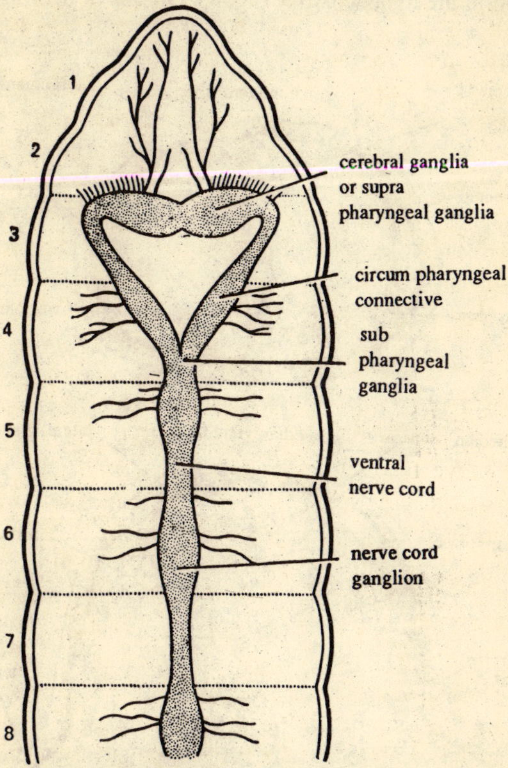

Fig 3.21. *Pheretima* Central System.

3. *Autonomic Nervous System.* It comprises an extensive plexus of nerve fibres present the lining epithelium and the circular muscles of the alimentary canal. The plexus is connected by fibres with the circumpharyngeal connectives.

Sense organs

As in other matazons, the organs concerned with the reception of stimuli from the environment are situated on or near outside of the body. These are called receptor organs or sense organs. These organs are quite simple in structure in the earthworms, consisting of either single or of small group of specialized extodermal cells. Those lying in groups forming the textile, olfactory and taste organs, while those occurring single form the light-sensitive organs. Organs of hearing are altogether absent.

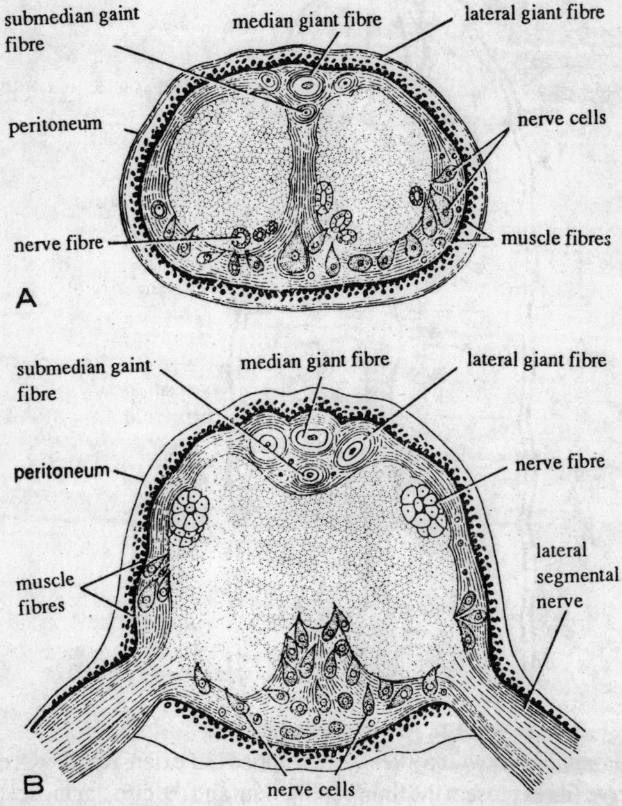

Fig 3.22 A-Section of ventral nerve cord, B-A section of the ventral nerve cord at the region of the origin of segemental nerves.

These are three types of receptor organs in *Pheretima.*

1. Epidermal receptor
2. Buccal receptors
3. Photoreceptors

1. *Epidermal receptors*. These are distributed all over the epidermis but are more abundant on the lateral sides and ventral surface of the body. Each receptor has an elevated cuticle covering a group of tall, slender and columnar *receptor cells,* bearing small hair like processes at their outer ends and connected with nerve fibres at their inner ends. They are surrounded on all sides by ordinary *supporting epidermal cells*. They are separated from each other by spaces and have nuclei at different levels and posses internally a few *basal cells*. They are *tactile* (relating to touch) in function and according to some, they also respond to chemical stimuli (*chaemoreceptors)* and changes in temperature (*thermorceptos*).

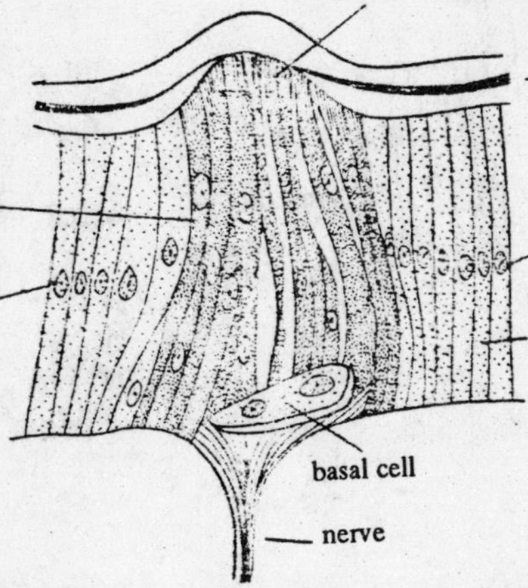

Fig 3.23. *Pheretima*. A buccal receptor

2. *Buccal receptors*. These are distributed in the epithelium of buccal cavity in large number. They consist of group of tall cells which project beyond the epithelial cells. They have sensory hair-like processes and their nucli lie below the middle part. These receptors serve to smell (*olfactorecptors*) and also taste food (*gustoreceptors*).

3. *Photo-receptors*. These organs are found in large number especially on the prosotomium and on some of the anterior segments. They are not fond in clitellum and on the ventral surface. However, they are a few in posterior region. Each organ consists

of a single ovoid cell or optic cell situated deep in the epidermis. There is a nucleus and clear cytoplasm containing net work of neurofibrilae. There is also a small transparent L-shaped lens or phaeosome which is somewhat tilted. Lens is formed due to a kind of transparent fluid. A few nerve-fibres are also connected with these cells. Lens severs to focus light and nerve-fibrils act as retina.

Photoreceptors are sensitive to light, but earthworms act negatively to all very weak light, hense they retreat into their burrows during day and emerge at night.

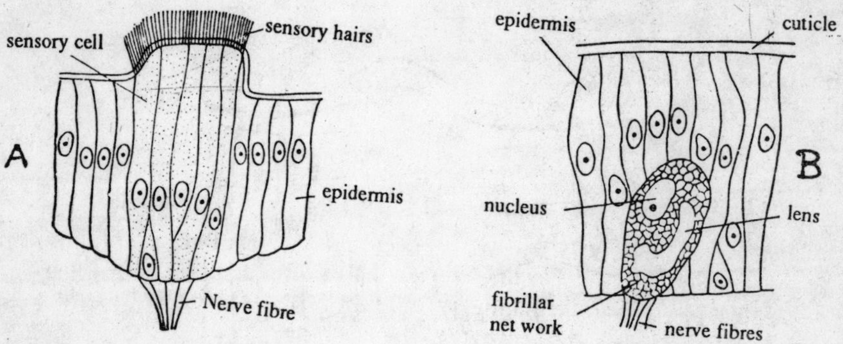

Fig 3.24. *Pheretima* A = Buccal receptor B = Photo receptor.

Reproductive System

The earthworms are *monoecious* or *hermaphrodite*. The reproductive organs are restricted to a few anterior segments. Though hermaphroditic, the self-fertilization not occurs in earthworm, but pair together (copulate) and later develop deposit eggs in cocoons. Moreover, the testes ripen earlier (protanday), make cross-fertilization necessary. Thus, hermaphroditism seems to be of no use to the earthworm.

Male Reproductive Organs. The male reproductive system in formed of testes, seminal vesicles, spermiducal funnel, vasa deerentia prostate glands and accessory glands.

Testes. There are two pairs of testes which produce sperms. There are minute, white and lobed structures. One pair is situated in the 10th segment and the other in the 11th segment on either side of nerve cord. Each testis is made up of a compact narrow base, from which arise 4-8 finger like processes containing spermatogonia.

Each pair of testes is enclosed in a thin-walled, fluid filled bag, the *testies-sac*. Thus there are two testis-sacs situated ventrolaterally, one behind the other, in segments 10th and 11th respectively. The poserior testies sac is larger than the anterior one.

There are two pair of spermatic funnel or spermiducal funnel in the two pairs of testies sacs. Each funnel is a ciliated structure and lies immediately behind its own testis and opens into a ciliated *sperm duct* or *vast deferens*.

Seminal vesicles. There are two pairs of large, white seminal vesicles one pair in each 11th and 12th segments respectively. The testis sac of 10th segment communicates with 11th segment and the testis sac of 11th segment with the 12th segment. The seminal vesicle of 11th segment is not visible as they remain enclosed in the testis sac of that segment. However, the seminal vesicles of 12th segment lie free.

Vasa deferentia. Each spermiducal funel leads postericaly into a fine ciliated duct, the vast deferens. The vasa deferential of the same side travel come together and meet a short prostatic duct of that side in the 18th segment.

Prostate glands. There is a pair of large, flat, white irregular organs situated one on either side of the intestine from 16th or 17th upto 20th or 21st segment,m each prostate gland gives off a short duct on its inner side. The prostatic duct is joined by two vasa deferential, thus, forming three ducts are called *common prostatic* and *spermatic ducts* and open to the outside ventrally on the 18th segment by the male genital pore. The opening of the male genital pore is actually a group of three apertures, a large prostatic aperture and two small apertures of vasa deferential. The function of the vasa deferential is to convey the seminal fluid to the outside for fertilization after copulation.

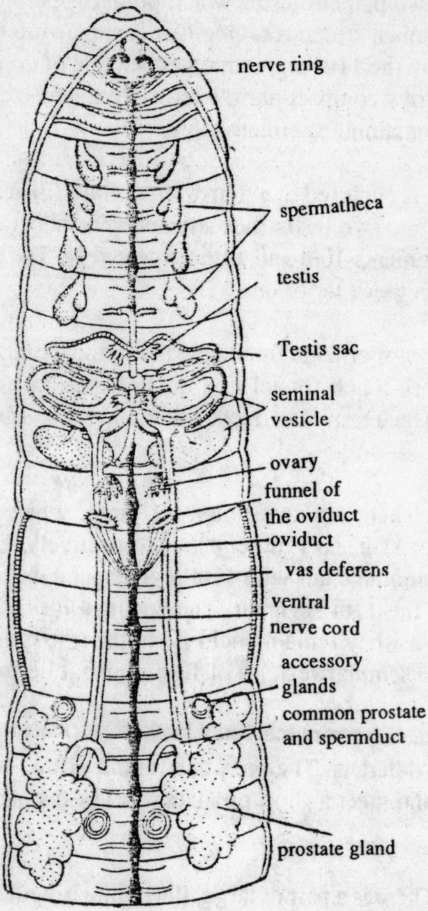

Fig 3.25 *Pheretima. Reproductive System.*

Accessory glands. The accessory glands are two pairs of rounded masses lying in the 17th and 19th segments just internal to the prostate glands. They open to exterior by a number of ducts on the ginatal papillae situated on the 17th and 19th segments. Their secretion helps probably in uniting the two worms during copulation.

From the testes the spermatogonia are shed into the testies sacs, from where thy pass into the seminal vesicles where they develop into spermatozoa by an elaborate process of spermatogenesis. Each sperma

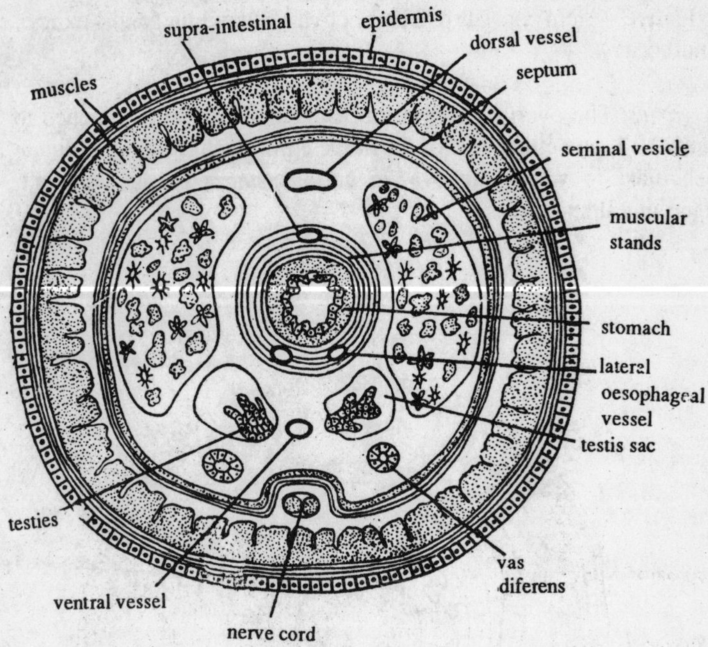

Fig 3.26 *Pheretima*. T.S. Through Seminal vesicle

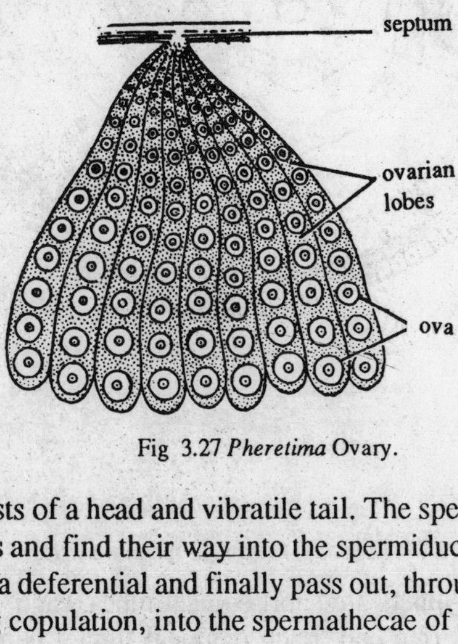

Fig 3.27 *Pheretima* Ovary.

tozoon consists of a head and vibratile tail. The sperms come back into the testis sacs and find their way into the spermiducal funnel and travel along the vasa deferential and finally pass out, through the male genital prores during copulation, into the spermathecae of another worm.

Female Reproductive Organs. The essential organs of the female reproductive system consists of a pair of ovaries,d oviduct and 4 pairs of spermathecae

Ovaries. The ovaries hang in the 13th segment, being attached to spetum 12/13, one either side of the nerve cord. Each ovary is a minute whitish mass in which the ova, in various stages of development, arranged in a linear series.

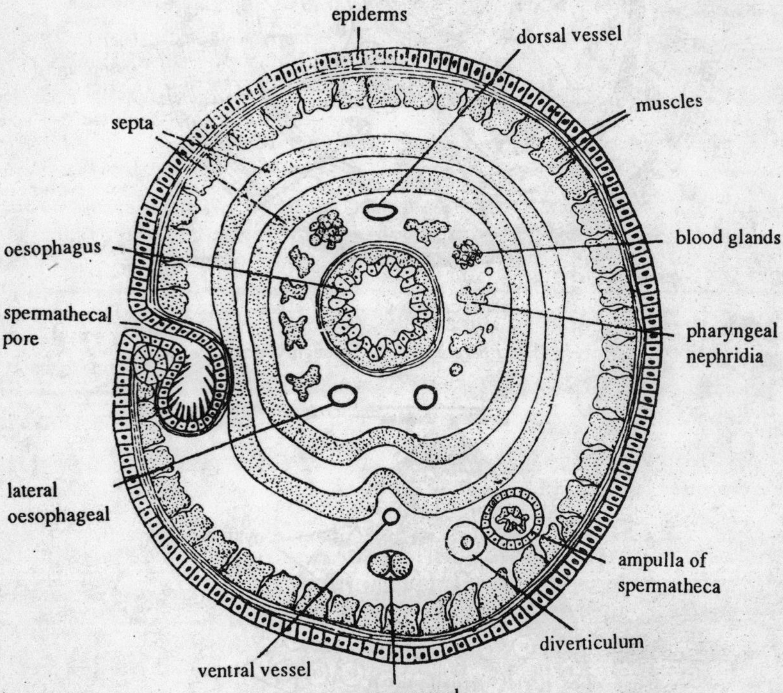

Fig 3.28. *Pheretima.* T. S. through spermechecal

Oviducts. The oviducts are short conical tubes, each with a large oviducal funnel at its free end. Each oviducal funnel is more or less saucer-shaped structure with folded and ciliated margins, and lies immediately behind the ovary. The oviducts run behind, perforate the septum 13/14 and coverage meting finally in the body wall beneath the nerve cord to form a short common duct that opens the *female genital opening* on the 14th segment. Internally the oviducts the oviduct are ciliated.

Spermatheace. There are four pairs of spermathecae situated in the 6th, 7th, 8th and 9th segments. Each spermatheca is a flask-shaped structure, the body is which is known as the *ampulla*. From the narrow neck arises a simple elongated *diverticulum* which stores the spermatozoa in *pheretima*. In other worms the spermatozoa are stored in ampulla.

Copulation

During rainy season when the whether is both warm and moist, the earthworms come out of their burrows at night and copulate in pairs. During copulation two earthworms become close applied to each other by their ventral surfaces, with the heads pointing to the opposite directions so that the male genital pore of one is opposite to the spermathecal pore of the other. The areas of male genital apertures are raised into papillae which are inserted successively into the various pairs of spermathecal pores from behind forwards thus filing the spermathecae with spermatozoa and prostatic fluid. After the completion of this process both the earthworms are separated and prepare for laying of eggs.

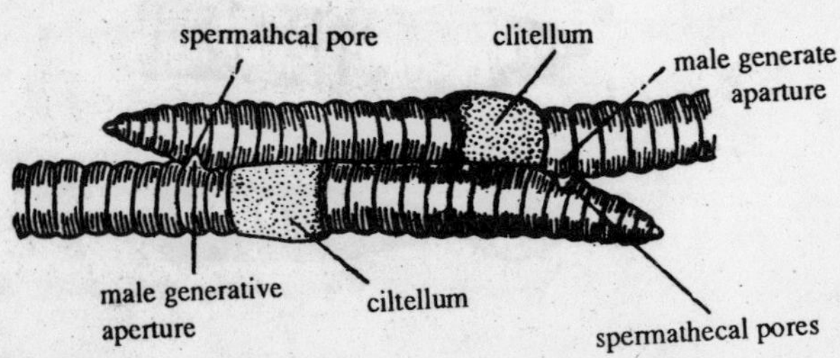

Fig 3.29 *Pheretima*. Showing Copulation

Fertilization

Cocoon formation is not well known is *Pheretima*, However, it has been studied in other worms like *Eisenia* and *Rhynchelmis etc*. The cocoons are formed usually during the months of August to October. If the condition are favourable the cocoons are also laid in April, May and June.

The cocoon is a product of clitellum. The clitellum is a girdle around the segments 14-16. It has swollen epidermis containing three types glands in three distinct layers, they are unicellular mucous glands which secrete mucous for copulation, cocoon-secreting glands secrete the wall of cocoon the albumen glands produce albumen in which the eggs are deposited in cocoon. The cocoon-secreting golands produce a viscid substance which forms an elastic membrandeous tube around the clitellum on coming in contract with air. The earthworm now starts moving backward so that this tube is pushed forward. The eggs come out of female genital pore are collected in it. By gradual backward movement of the worm, this tube reach over the spermathecal pores and the sperms are released from spermathecae. Thus the fertilization takes place is external. As a soon as the earthworm withdraws itself from the tube, the openings due to elasticity closed at both the ends and, becomes a small rounded capsule, called the cocoon. The cocoon this formed is left in the soil, it is light yellowish in colour and about 2-2.4 mm in length and 1.5-2 mm in width.

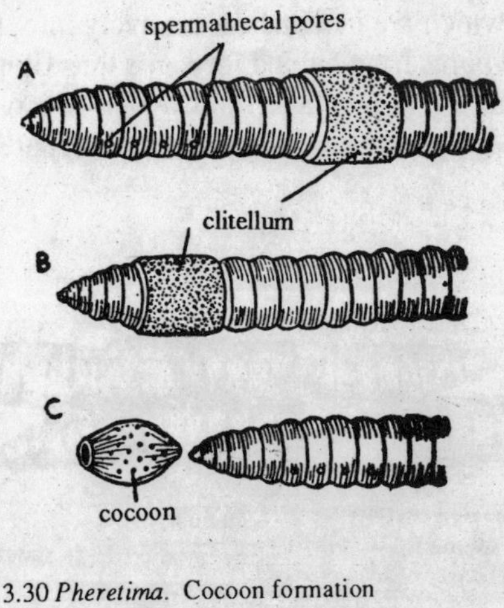

Fig 3.30 *Pheretima.* Cocoon formation

Development

The zygote contains very little amount of yolk. Cleavage is holoblastic but unequal. Two types of cells formed, the small one are called *micromeres* and the larger ones are called *meagameres*. Continued segmentation results in the formation of a hollow spherical body, the *blastula*. The walls of baltula are composed of single-layer of cells, but the ones at the lower end are larger than those above. Two cells lying beside each other near the equator become distinguishable from the rest by the fact that they are quiescent, whereas the rest continue to divided. These are the *mesoblastic pole cells* or pole-cell of mesoderm. After some times these pole cells divide repeatedly to give rise to two rows of small cells, sivide repeatedly to give rise to two rows of small cells, which, together with the pole cells constitute the mesoblastic bands.

Now the blastula elongates and flatens forming a structure with the small cells above and large columnar cells below, while the mesoblast cells lie on the equatorial plane.

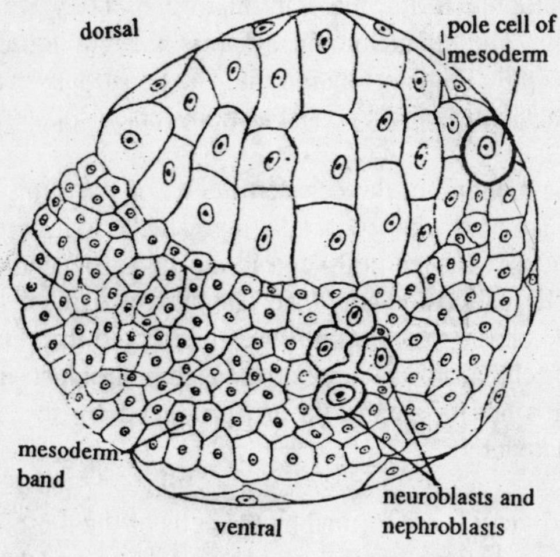

Fig 3.31 Embryonic development in earthworm. Ventro-lateral view of young embryo.

Gastrulation beings at this stage by invagination. The lower surface made up of megameres is pushed inwards. The micromeses multiply and keep on growing over the megameres. Due to inrushing a cavity is formed, the *archenteron and the groove,* produced due to in vagination is a *blastopore* on lower surface. With the completion of process of gastrulation, the three germs layers are established. The outer embryonic cells constitute the *ecoderm,* the cells lining the archenteron, the *endoderm* and the mestoblastic bands, the *mesoderm.*

The blastoporal groove begins to close from behind forwards till only a small opening is left at the anterior end. The persists as mouth and leads into a small *stomodaeum* lined by ciliated ectoderm. The stomodeaum leads into the archenteron. The anus is formed in a very advanced embryo as an ectodermal in vagination fusing with the enter on in the last segment of the worm.

The mesoblastic bands are situated parrel to each other on either side of mid-ventral line. Each band segments into a row of blocks, the mesoblastic somites, which are solid at first, but develop each, a cavity in the interior secondily. The lateral somites with their cavities grow dorsally and ventrall ultimately fusing with each other, with the result that a continuous space is formed in the mesoderm in contact with the body wall is the *somatoplure* and the inner layer is the *splanchnoplue*. The somatoplue forms the musculature of body wall and coelomic epithelium. The splanchnopluse forms the gut musculature and the coelomic epithelium surrounding it. The partitions between the successive mesoblastic somites persist as transverse septa.

In the meantime, the ectoderm cells of the ventral surface of the embryo become thickened and distinguished into four longitudinal rows of cells on each side of mid-ventral line, each row ending posteriorly in a large cells, called and *teloblast*. The two rows of cells immediately on either side on the mid-ventral line, give rise to the *ventral nerve cord,* hence the cells forming it are known as th neuroblast called nephroblast cells. All kinds of nephridia are produced by these cells but they differentiate later.

Blood-vessels are formed by the cells of the mesoderm, the ventral and sub-neural appearing in the mid-ventral line, and the dorsal at the top of the somites of each side. As somites of the two sides meet and fuse above, the paired dorsals fuse to give rise to the medium dorsal vessel.

Thus the ectoderm gives rise to the epidermis, nervous and excretory system as well as proctodaeum and stomiodeaum. Mesoderm gives rise to body musculature, coelomic epitheliuum and the blood vessels. The endoderm gives rise to the gut and its appendages, glands etc.

Further development of worm takes place by the formation of new mesoblastic somites from behind forwards till the embryonic development is completed.

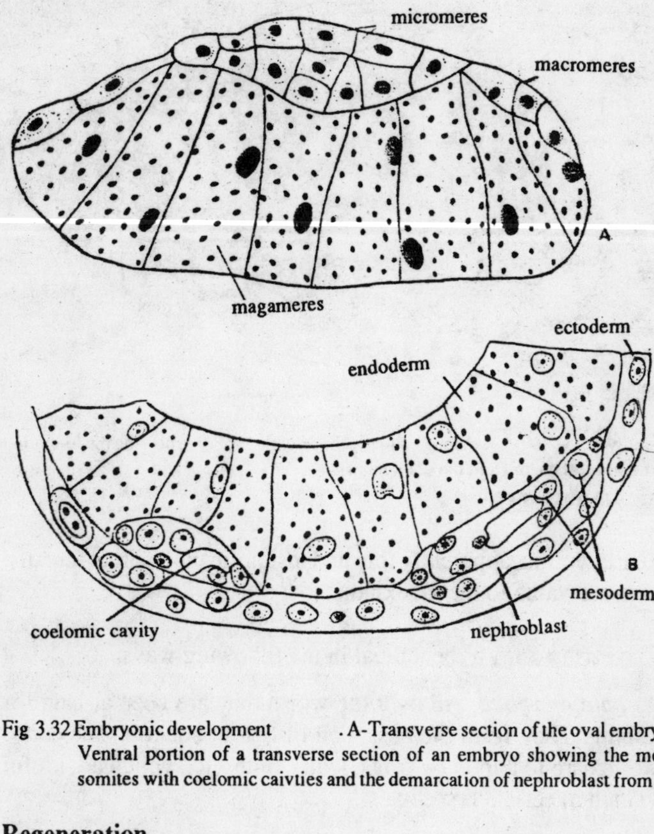

Fig 3.32 Embryonic development . A-Transverse section of the oval embryo; and B-Ventral portion of a transverse section of an embryo showing the mesoblastic somites with coelomic caivties and the demarcation of nephroblast from ectoderm

Regeneration

Earthworms posses the ability to regenerate lost part, but this ability is limited. A new head may be regenerated after the removal of upto 15 anterior segments, but the capacity to regenerate decreases rapidly behind the 9th segment. The posterior region are more easily regenerated. If the cut is made near the tail end, a new anterior end will not regenerate on the tail half, but instead another tail will develop from cut surface. When any portion of the tail region is cut off, the last parts readily regenerate. The number of segments regenerated is usually less then the number removed.

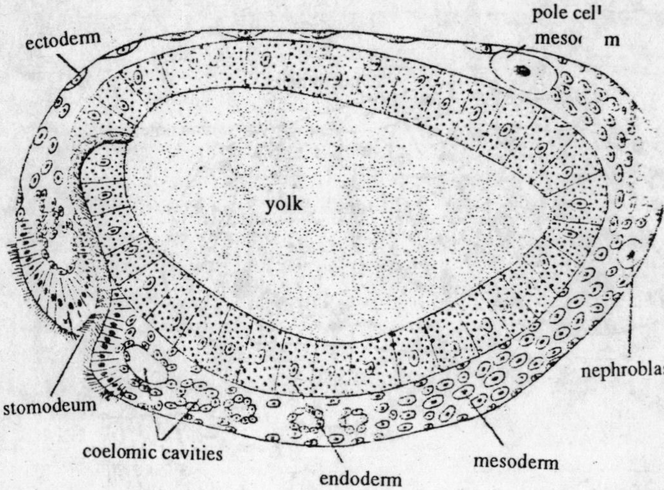

Fig 3. 33 Embryology . A longitudinal section of the embryo showing the formation of the stomodaeum and the coelomic spaces

Economic Importance

The earthworms, apparently tiny and insignificant creature, are of a great economic value to the mankind.

I. The earthworms ar beneficial in the following ways:

1. *As baid and food.* All over the world they are used at baid for fishing. They form the best food of fish in aquaria. The earthworms are fed upon by frogs, toads, birds etc., which are useful to man in certain respects.

2. *In agriculture.* Earthworms are in general beneficial to agriculture. They improve the soil. This is done in many ways. Their habit of burrowing and swallowing earth below the surface increases the fertility of the soil. The soil is made porous which allows better aeration, quick absorption of water and easily penetration of roots. Waste matter and faces thrown out by worms have a good deal a manurial value. The faces has more nitrate, calcium, phosphorus etc. They bring about optimum conditions for plant growth by reducing both acidity and alkalinity of the soil. A more important beneficial effect on soil is the mixing and churning of solid resulting from burrowing. Deeper soil is brought

to the surface as casting and organic material is moved to lower levels.

3. *In mediucines*. Earthworms var... ly ... medi...nes in the past to use stones of bladder, pyorrhoa, jaundice etc.

4. *In laboratories*. The earthworms is studied as a representative the Annelida by students of Biology.

II Harmful activities

Sometimes they destroy tender plands by eating them. Their burrows on the sloping land promote solid erosion during rainy season. The spoil golf grounds.

4

HIRUNDINARIA GRANULOSA

Class Hirudinea (Lat. *Hirudo*=a leech) form a relatively small but well-defined group of segmented wooms which are generally adapted to an ectoparasitic mode of life. These are commonly called 'leeches' About 300 species of leeches are known. They abound in tropical and sub-tropical regions. Most of them are freshwater, while some are marine or terrestrial. Majority are blood suckers and parasitic, others are predators. feeding on worms, snails, insect larvae etc.

Genus *Hirudinaria* includes four indian species namely, viridis, *H. manilensis, H. javanica* and *H. granulosa. H. granulosa* is the more common species and comonly known as cattle leech.

SYSTEMATIC POSITION

Phylum	-	Annelida
Class	-	Hirudinea
Order	-	Gnathobdella
Genus	-	*Hirudinaria*
Species	-	*granulosa*

Habits and Habitat

Hirudinaria is a comon Indian leech found in India, Burma, Sri Lanka, Bangla Desh and Pakistan, It occurs in freshwater ponds, lakes, slow streams etc., where it either swims by vertical undulations or grips objects with its suckers and moves by looping. It is *sanguivorous* in diet e.i. feeds on the blood. A single full meal can last for more than a year.

Leech is bisexual but always cross-fertilization occures. Cross fertilization involves pairing of two individuals in head-to-tail position. Eggs are deposited in cocoons, Development is direct i.e. without any free larval stage in the life history.

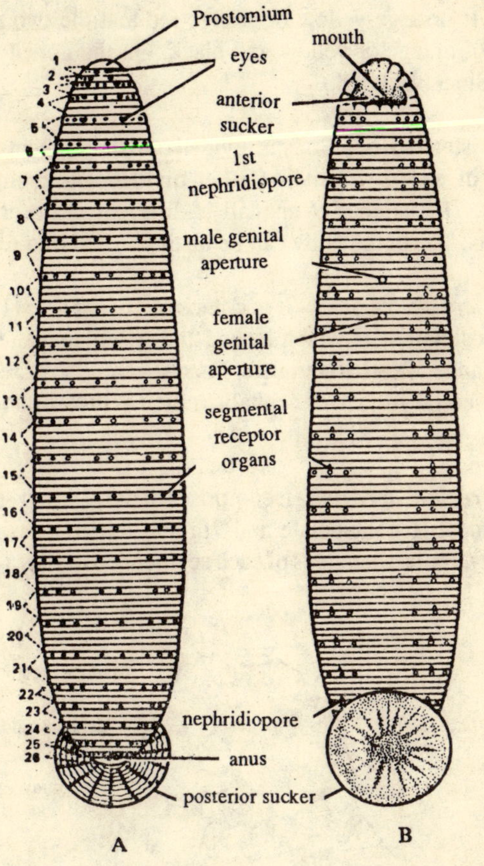

Fig 4.1 *Hirudinaria granulosa.* (A) Dorsal view (B) Ventral View

External Morphology

Shape and size Body is elongated dorsoventrally flattened and vermiform with flat ventral and convex and dorsal surfaces. A fully stretched leach appears to be ribbon-shaped, whereas in contracted condition it is cylindirical. Often if measures about 4-6" but might reach upto 12" or more.

Colouration. The body is beautifully coloured with characteristc markings. The dorsal surface is generally olive-green and the ventral surface is orrange-yellow or orange-red and the two sides bear distict stripes of orange or yellow and black. On the dorsal side is a median longitudinal black strip.

Division of Body Body is divided into 33 segments but the external surface of each segment is further divided into annuli by transverse grooves. The number of annuali is different in different in different segments. The entire body can be distinguished into 6 regions.

1. *Cephalic regions* - It is comosed of first five (1st-5st) segments and a prostomium, of which 1st and 2nd are uniannulate, 3rd biannulate, and 4th ande 5th are triannulate. Each segment possesses a pair of eyes. On the ventral surface of cephalic region is present anterior sucker and mouth.

2. *Preclitellar region* - It comprises of 6th, 7th and 8th segments. The 6th segment is traiannulate and 7th and 8th consists of five annuali. Ventrally, the last annualus of each segment possesses a pair of nephridiopores.

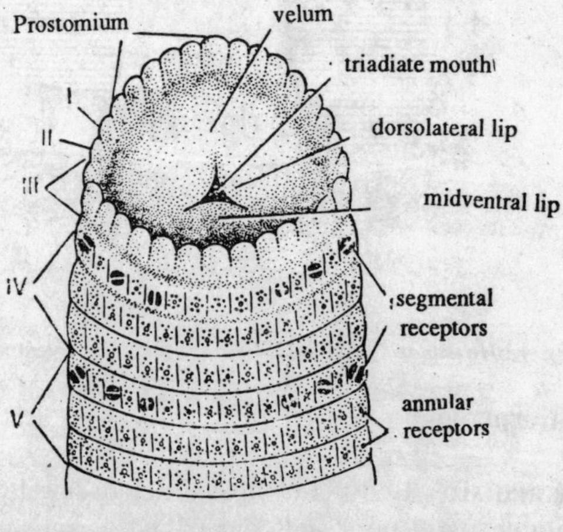

Fig 4.2 *Hirudinaria*. Cephalic region in ventral view.

3. *Clitellar region* - It extends from 9th to 11th segments. Each segment consists of five annuali and carries a pair of nephridiopores. Clitellum is distnct only during reproductive season,

4. *Middle region* - It extends from 12nd to 22nd segment. All the three segments are quinquannulate (5 annulate) and possess nephridiopores.

5. *Caudal region* - The caudal region comprises of three segments from 23rd to 26th. These are without nephridiopores. Segment 23 is triannulate whereas the rest are *biannulate. Anus* is located middorsally in 26th segment.

6. *Posterior sucker region* - Posterior sucker is formed by seven segments from 27th to 33rd. These are uniannulate and are arranged in concentric rings.

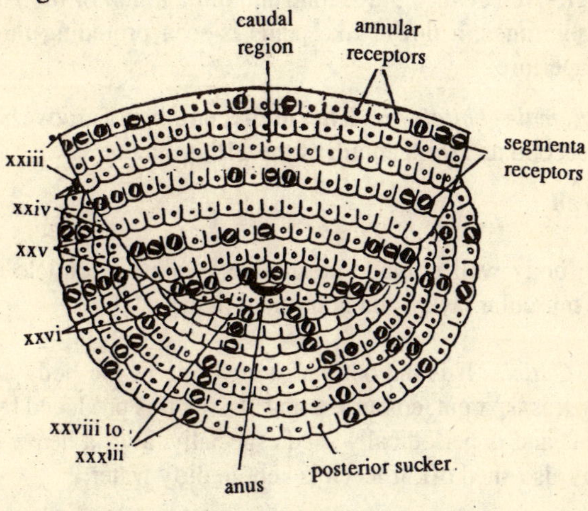

Fig 4.3 *Hirudinaria* Caudal region in ventral view

Suckers. Each end of the body bears a hollow muscular organ, the *sucker*. The anterior or *cephallic sucker* comprises the prostomium and three anterior body-somites. It is oval in outline with a ventrally directed, cup-like hollow, called athe pre-oral chamber, at the bottom of which lies the mouth. The posterior end also bears a circular and highly muscular disc, the posterior or anal sucker, also directed downwards. It is formed by the complete fusion of the last seven body segments Both the suckers are primarily meant for adhesion and locomotion. The anterior sucker also helps in feeding.

Apertures. The body of leech has a number of apertures. These are as follow:

(i) *Mouth.* It is a narrow triradiate aperture situated in the centre of the preoral chamber of anterior sucker.

(ii) Anus . It is a small aperture situated mid-dorsally on the 26th segment at the junction of the body and posterior sucker.

(iii) *Nephridiopores.* - 17 pairs of minute nephridi pores present on the ventral side of the body, a pair on the last annulus of each of the segments 6-22.

(iv) *Male genital pore.* It is a mid-ventral opening, situated in a groove between the second and third annuli of the 10th somite. somtimes, a thread like penis is seen protuding through, this aperture.

(v) *Female genital pore.* It lies mid-ventrally in a groove between the second and third annuli of the 11th segment.

Body-wall

The body wall is made up of five layers viz, cuticle epidermis, dermis, muscular layers and botryoidal tissue.

1. *Cuticle* It is the outermost covering of the body and is thin, delicate, trasnsparent, colourless and elastic. It is produced by the inner epidermis and is periodically shed especially, after a heavy meal. The cuticle is also shed off, if leech is kept in dirty water.

2. *Epidermis.* Lying below the cutcle is the *epidermis,* which consists of a single layer of hammer-shaped cells. These cells are broad, flatened and pentagonal towareds and outer, and narrower towards their inner ends. The outer ends fit closely while the nucleated inner ends have interspaces containing fibres connective tisse, pigment cells and haemocloeomic capillaries, forming a vascular and respiratory membrane.

These are several glands found in the epidermis. These are modifications of epidermal cells.

(i) *Slime-glands* are distributed all over the body and produce a kind of slimy mucus, which covers whole of the body. The tubular body of the slime-gland reaches deeper into dermis and possesses a nucleus, a thin layer of cytoplasm and several secretory granules. Through the narrow neck, each gland opens to the outside by a minute a perture in the cutcle.

(ii) *Sucker glands* -These are also pear-shaped or rounded in form and occur in clusters in both the suckers. Their secretion serves to smoothen the surface for the attachment of suckers during locomotion.

(iii) *Prostomial glands* - These occur in groups in the prostomium. Their secretion plugs the ends of ootheca or egg-case.

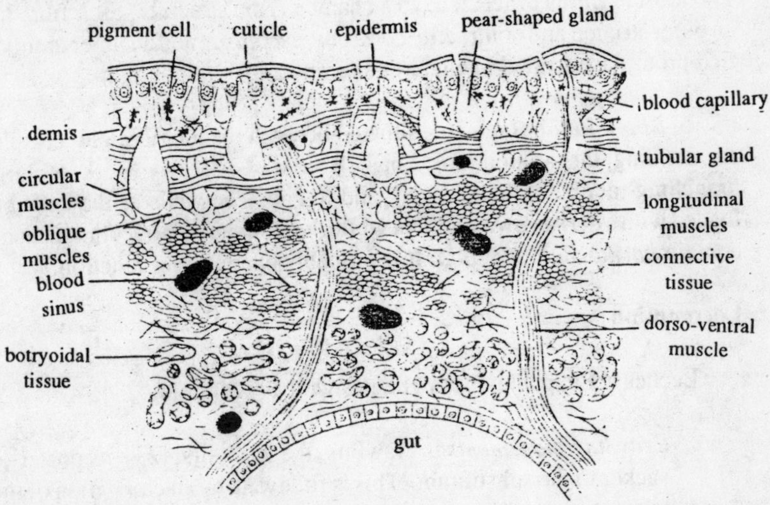

Fig 4.4 *Hirudinaria* Section of body wall.

(iv) The *Citellar glands* are found in the clitellar region (segment 9 to 11) and become active during the breeding season. They are of two special types, the chitogenous glands forming the wasll of the cocoon, and the albumen gland secreting albumen that fills up the cocoon. The former lie between circular and oblique muscels and the latter in longitudinal muscles.

3.*Dermis*. It lies under the epidermis. It also extends into the spaces left between the inner narrow ends of the epidermal cells. It is made of connective tissue and consists of gelatinous matrix containing interlacing fibres and cells. It also contains pigments and fat cells, scattered muscle fibres and branches of haemocoelomic channles.

4. *Museles* The musculature lies beneath the dermis and consists of thin layer of *circular muscles* and a thicker layer of *longitudinal muscles.* The longitudinal muscels are powerfully developed and their fibres converge into the two suckers. Circular muscles are arranged concentrically in the the suckers. Between the circular and longitudinal muscles leeches have an additional double layer of *oblique muscles* directed spirally around the body like a coil. There are also dorsoventral muscles arranged segmentally thoughout the body, they run from the dorsal to the ventral side in each segment. There are some radial muscles whose fibres radiate from the alimentary canal to the skin, they take the place of septa. The muscle fibres of leech have a characteristic structure, each fibre has an outer striated and contractile *cortex* or *myoplasm*, and an inner unmodified protoplasm called *medulla* or *sarcoplasm.*

5. *Botryoidal tissue* - It is situated beneath the longitudinal muscles surrounding the alimentrary canal. It consists of a network of large branching tubular cells arranged end to end. The walls of the cells are loaded with a dark brown pigment while their lumen or intracellular canals contain *haemocoelomic fluid*. It is probably excretory in function.

Locomotion

Leches are capable of crawling as well as swimming.

1. *Crawling movements* In crawling, the leech first fixes its posterior sucker to the substratum. This is followed by an elongation of the body on the substractum brought about by the contraction of the circular muscles and relaxation of the longitudinal muscels. The animal then fixes its anterior sucker to the substratum and draws fowards its posterior sucker and the rest of the body by the contraction of the longitudinal muscles and relaxation of the circular muscels. The posterior sucker is then fixed to the substratum close behind the anterior sucker, so that the body forms a loop. The posterior sucker now remains fixed and the rest of the body elongates by the mechanism described above.

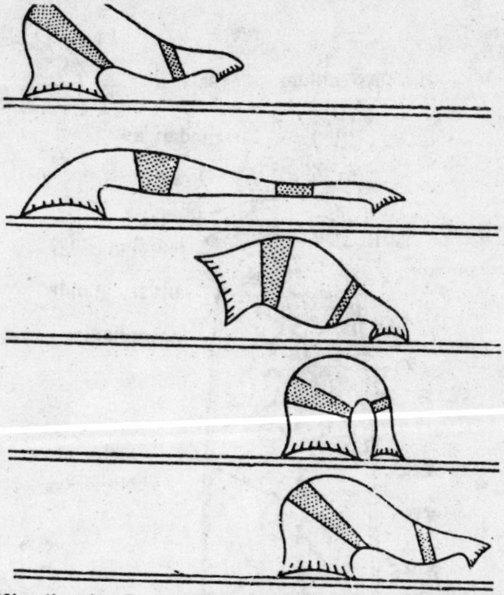

Fig 4.5 *Hirudinaria.* Stage of locomotion.

2. *Swimming* It is brought about by the undulatory motion of the body, which is in a state of dorso-ventral flatenning owing to the sustained contraction of dorso-ventral muscles. The undulatory movement is brought about by waves of contraction and relaxation passing along the longitudinal muscles.

Digestive system

The digestive system comprises the alimentary canal and glands associated with it.

Alimentary canal. It is a straight tube of varying diameter running from the anterior to the posterior end. It is supported on all the sides by botryodial tissue. It is differentiated into mouth, buccal cavity, pharynx, oesophagus, crop, stomach, intestine and rectum.

1. *Preoral chamber*-The cup-shaped depression of the anterior sucker represents the preoral chamber. Its base is perforated by triradiate *mouth* which opens itno the buccal cavity. The preoral chamber and buccal cavity are separated from one another by a membranous partition, the *velum.* It consists of three flaps, which form three lips of the mouth.
2. *Buccal cavity.* The mouth leads into the *buccal cavity,* which is a short chamber behind the velum, In its mucous lining are

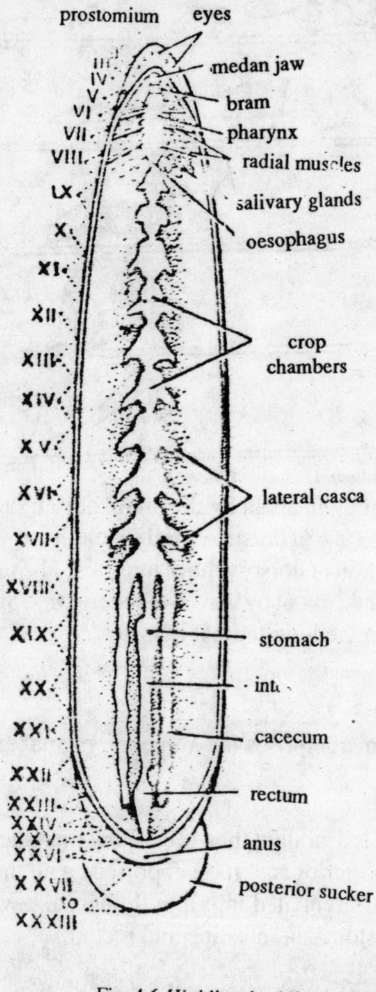

Fig 4.6 *Hinidinaria*. Alimentary Canel.

embedded three cresentic jaws, each lodged in a depression. One jaw lies behind each ray or chink of the mouth, i.e. one is mediodorsal and the other two ventro-lateral in position. Each jaw is a laterally compressed muscular cushion, covered with fine cuticle which is thickened at the free edge to form a ridge bearing minute *teeth* or *denticles* in a single row. Such jaws are termed *monostichodont*. The number of the denticles is 103-128 on the median jaw and 85-115 on lateral jaw. On each side of the jaw are about 42-45 buton-shaped protuberances, the *salivary papillae*, each bearing numerous opening of the salivary glands.

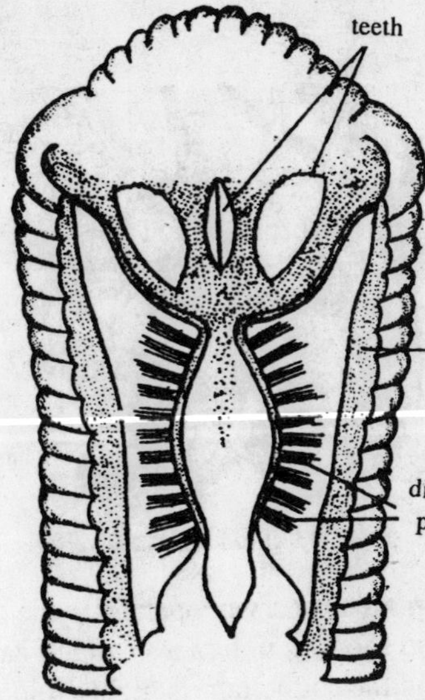

Fig 4.7 *Hirudinaria.* Ventral dissection of anterior end showing there jaws

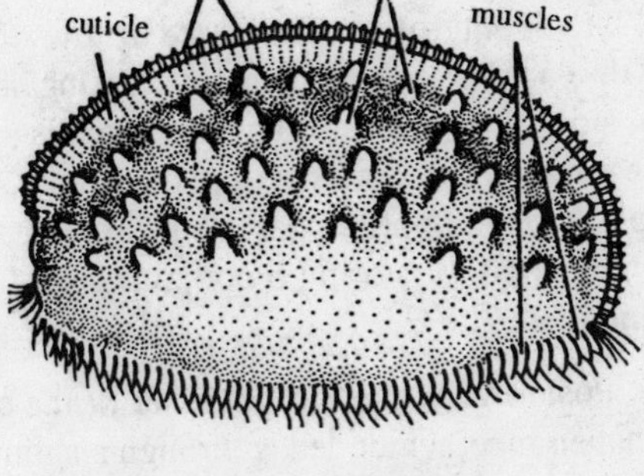

Fig 4.8 *Hirudinaria.* A jaw

3. *Pharynx.* The buccal cavity is followed by a highly muscular, oval sac, the pharynx. It extends from 5th to 8th segment (Fig 39.11) It is covered with unicellular salivary glands, whose ducts open into the salivary papillae of jaws. The entire spoace between the wall and the pharynx is filled up by salivary glands. The saliva secreted by salivary glands has a kind of substance . u . *hirudin* or *anticoagulin,* which does not allow the blood to coagulate during the act of sucking it from the prey.

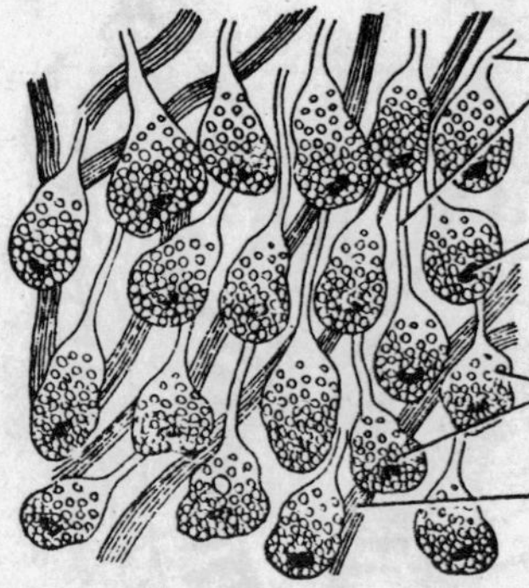

Fig 4.9 Hirudinaria. Salivary glands

4. *Oesophagus* The pharynx opens into the oesophagus, which is a narrow and short section of alimentary canal. The lumen is very narrow and the epithelium is highly folded.

5. *Crop :* The crop is the largest region of the alimentary canal extending from the ninth to the eighteenth segment. It has ten thi-walled chambers, one in each segment. Each chamber consists of a small anterior and a broad posterior part which is produced into a pair of lateral outgrowth. The *caeca or diverticula* the chambers and the caeca go on gradualluy increasing in size towards the posterior side., The last chamber of the crop is the largest and its caeca are prolonged backwards upto the twenty second segment. The crop is a very important part of the alimentary canal, as it is elastic and is used for storing blood, one crop-ful of blood lasts several months.

6. *Stomach.* Posterirorly, the last chamber of the crop ends into a funnel-shaped tube, which leads through a sphinctered aperture into a small heart-shaped *stomach,* lying in the *19th segment.* The mucous lining of the stomach is thrown into anastomosing *transverse folds.*

7. *Intestine* -It is a thin-walled narrow tube from 20th to 22nd segment. Its lining is also produced into spiral villi-like processes to increase the surface of absorption. The stomach and intestine are not marked externally.

8. *Rectum* - Behind the intestine lies the thin-walled wide rectum which extends upto 23red-26th segment. There are no internal

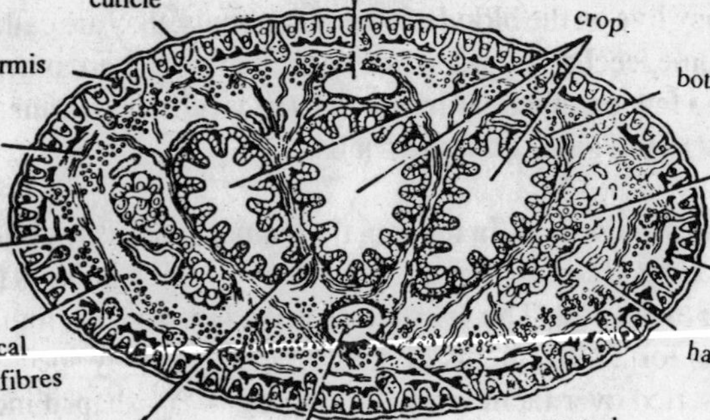

Fig 4.10 *Hiridunaria*. T.S. body through crop and its diverticula.

foldings in rectum. It opens out posteriorly by an *anus* which lies on the dorsal surface of 26th segment.

Histology

The pre-oral and buccal cavity are lined with cuticle continous with that of external surface. They form the stomodaeum. The rectum is also lined with with cuticle and forms the proctodaeum. The remaining parts of alimentary canal are lined with endoderm and forms the midgut.

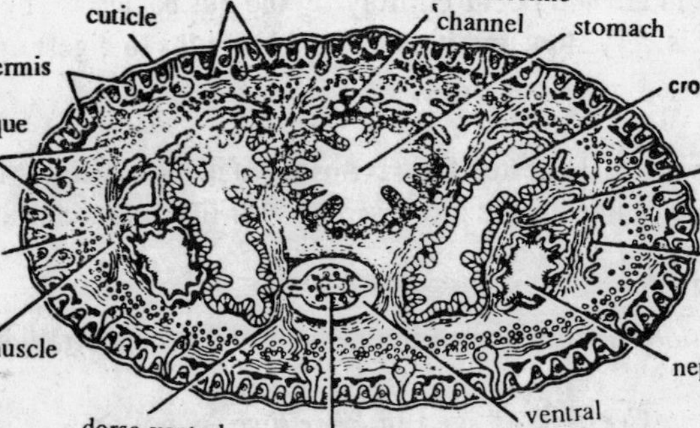

Fig 4.11 *Hirudinaria*. T.S. body through the region of stomach.

Food. Food of these highly organised animals is usually blood and, because, they live on the blood of various animals, they are called blood-suckers. Some leeches are parasites on the blood of man, others live on horses and a few on other cattle. some large leeches in Northern United States, feed voraciously on large snails.

Feeding or ingestion. In feeding the animal attaches itself to its prey by the posterior sucker. At the other end the pre-oral chamber is flatenned and applied to the host. The jaws are protruted through the mouth. Soon the back and forth movements of jaws brought about by special sets of, muscles inserted over them work a triradiate or 'Y'-shaped incision in the skin of victim. The incision lays open the cutaneous vessles from which the blood is sucked with the help of pre-oral chamber, the buccal cavity and the pharynx that form highly efficient apparatus for the purpose. A leech can suck blood several times its own weight in a single meal. When the crops are full of blood it drops of and preers darker places and remains hidden till this meal exhausts. The salivary glands produce a substances,, *hirudi*' which prevents the coagulation of blood during their feeding time.

Digestion. In the crops the blood is haemolyzed i.e. the red blood corpuscles burst and give up their haemoglobin into the plasma. Some water is absorbed in the thin walls of the crops. Digestion of haemolyzed blood takes place in the stomach by the proteolytic enzymes (*Abderhalden* and *Heise,* (1909). *Busing* and his colleagues discovered in 1953 that digestion in *Hirudo* occur entirely by the gut bacteria. The process of digestion is very slow and it takes 10-14 months to digest one crop-full of blood.

Absorption. Digested food is absorbed in the intestine. The abosrp-tive surface of intestine is increased by villi-like transverse and longitu-dinal folds.

Egestion Elemination of indigested food occurs through the anus.

Coelom. Coelom, as seen in *Pheritima,* is not present arround the alimentary canal of the leech Actually, the coelom is present, although in a considerably reduced state, owing to the growth of botryoidal tissue. It is represented by the four longitudinal vessels, which are not true blood vessels, but shut off parts of the coelom, the branches of these vessels and a fairly large number of spaces.

The *coelomic fluid,* running through the channels and their branches, contains colourless amoeboid *corpuscles* and is coloured red due to haemoglobin dissolved in it. The blood-like coelomic fluid is called the haemocoelomic fluid and the channels are termed the haemocoelomic channels, which together form the *haemocoelomic system.*

Thus, in *Hirudinaria, there are not true blood vessels, their function being taken up by the haemocoelomic system.*

Halmocoelomic system it consits of following four longitudinal vessels:

1. A dorsal longitudinal channel

2. A ventral longitudinal channel

3. Two lateral longitudinal channels

1. *Dorsal Haemocoelomic Channel* It is thin-walled non-contractile vessel running middorss ally above the alimentary canal from one end of the body to the other. The haemocoelomic fluid flows from behind forward. The valves are absent. Anteriorily, it remifies in the 6th segment and supplies the anterior five segments. Posteriorly, it bifurcates in the 22nd segment; the two branches run ventrally around the rectum to join ventral sinus.

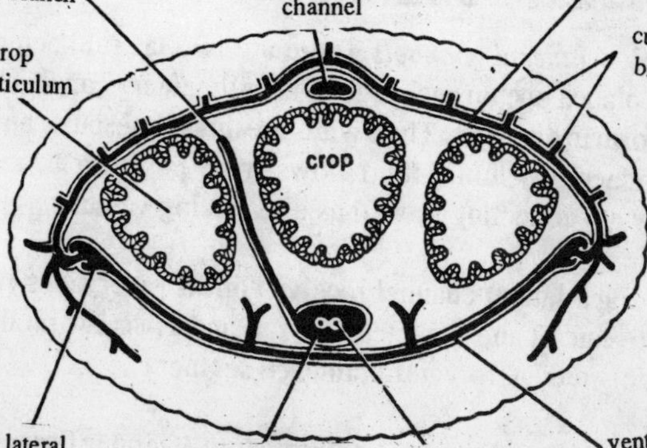

Fig 4.12 *Hinudinaria.* T.S. body through dorsal and ventral commissures of the lateral haemocoelomic channels.

It gives two types of branches :

(i) *Dorso-laterals.* Two pairs of *Dorsolateral branches* arise from the ventro-lateral side of the dorsal channel in each segment. Each branch runs outwards to form a capillary plexus in the dorsals and dorso-lateral regions of the body wall.

(ii) *Dorso-intestinals* These are numerous small branches arising from the ventral side of the dorsal channels, all along its length. They supply the gut wall.

2. *Ventral Haemocoelomic channel*. It runs below the alimentary canal in the mid ventral region from anterior to the posterior end of the body. The flow of the blood is from anterior to the posterior end. It encloses the nerve-cord, cerebral ganglia, nerve-ring and sub-pharyngeal ganglion. It is wider than the dorsal sinus but there are no valves. All the arches are segmental.

(i) First pair of branches originate at the lavel of nerve-ganglion and give rise to two main branches: (a) *Ventral branches,* which is found on the ventro-lateral side of the body-wall. (b) **Abdomino-** *dorsal* branch which runs upwards between the diverticula of crop and supplies dorsolateral part of the body-wall.

(ii) The seond pair is of nephridial branches, which are found in ii segments i.e. from 12th to 22nd segment, having testis-sacts. Each branches forms 2 or 3 sets of saccules, the *perinephrostomal ampullas,* in which ciliated organs are enclosed and fanally, passes out and gives off branches to lateral part of the body wall.

3. *Lateral channels* There are two lateral haemocoelomic channels placed one on each side of the alimentary canal. They are wide and uniform in diameter. Their walls are thick,, muscular and possess valves. The haemocoelomic fluid flows from posterior to anterior direction. They are collecting as well as distributing vessels.

Each lateral channel receives on its outer sides two branches, the latero-lateral and the latero-dorsal, and gives off on the inner side one branch, the latero-ventral, in each segment.

(i) *Latero-lateral.* The short latero-lateral is formed by the branches from the lateral region of the body wall and the nephridia. It joins the lateral channel at the level of the nephridial vesicle.

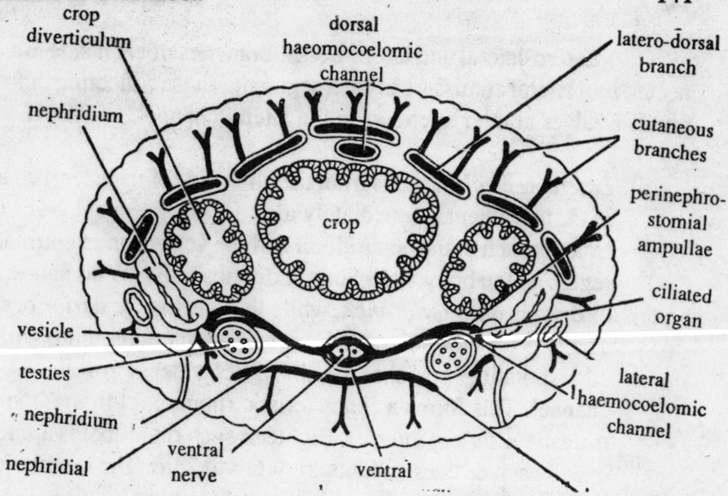

Fig 4.13 *Hirudinaria*. T.S. body through nephridial branches.

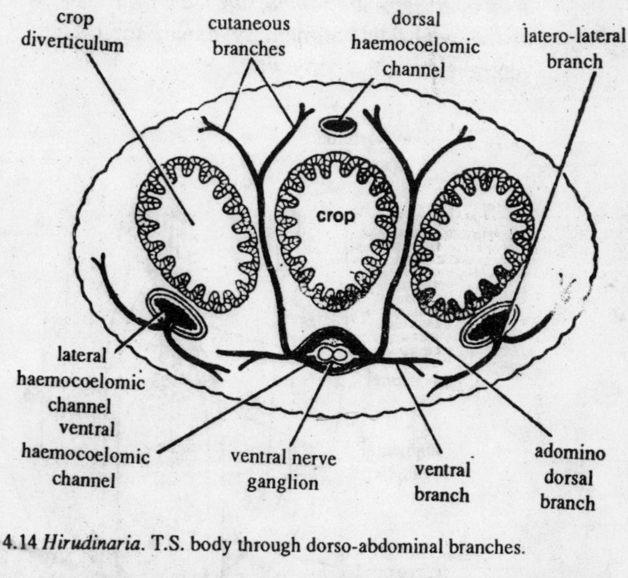

Fig 4.14 *Hirudinaria*. T.S. body through dorso-abdominal branches.

(ii) *The latero-dorsal* It opens into the lateral vessel at the level of nephridium proper, This is a large branch and collects blood from the dorsal and dorso-cateral branches of both the sides are connected by a dorsal commissure, which runs transversely above the dorsal canal in the segments from 6th to 22nd.

Latero-lateral and latero-dorsal branches of each side are connected by a lateral comissure in each segment. These collecting branches possess valves at their openings in the lateral channel.

(iii) *Latero-ventral* It arises from the inner side of the lateral channel in each segment. Immediately after its emergence it gives off a small branch to the nephridium and the ventral and ventrolateral region of the body wall. Soon it divides into two branches. One of them is *anterior branch*, while the other is *posterior branch.* The two branches diverge from one another and unite with the corresponding branches of the other side below the ventral channel. This forms a characteristic rhomoboid figure, the *ventral commisure of laterals*. Eighteen such rhomoboids are found one in each of the segments from 6th to 23rd. The rhomboids of the adjacent segments are also joined together by three *longitudinal intersegmental commissures.* the later-vantrals issue branches to the nephridia, alimentary canal, reproductive organs and ventral side of the body wall.

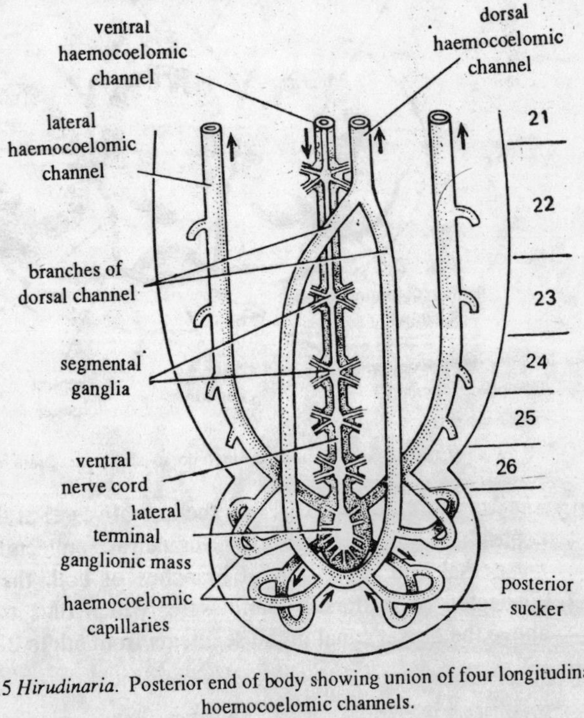

Fig 4.15 *Hirudinaria.* Posterior end of body showing union of four longitudinal hoemocoelomic channels.

Course of circulation

Of the four longitudinal channels, the fluid in the dorsal and the two lateral channels move in anterior direction, while in ventral channel it moves in posterior direction. The dorsal and ventral are the distributing channles, while the laterals are collecting as well as distributing chan

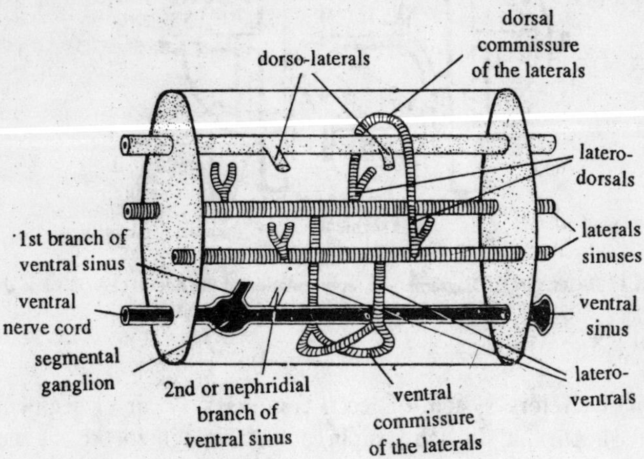

Fig 4.16 *Hirudinaria.* Showing important Channels

nels. All the channels are in direct communication in with one another at the posterior end i.e. in 26th segment. The dorsal and ventral channelos supply the haemocoelomic fluid to the body wall and dorsal surface of the alimentary canal. The ventral canal supplies the ventral and ventro lateral parts of the body wall as well as the nephridia. The latero-lateral and latero-dorsal branches of the lateral channels collect the fluid from these parts. The latero ventral branches supply the nephridia wall of aliment ary canal, ventral body wall and reproductive organs.

Respiratory System.

Leeches are devoid of special respiratory organs. The skin which has a rich supply of the haemocoelomic fluid act as respiratory surface for the exchange of gases. The epidermis acts as a permeable membrane through which the carbon dioxide from the haemocoelomic fluid diffuses out. Oxygen which is dissolved in water passess in the haemocoelomic fluid. For this purpose the skin is always kept moist and a copious secretion of mucous prevent it from drying. Some leeches such as *Branchellion* possess lateral out-growths of the body wall that actas gills.

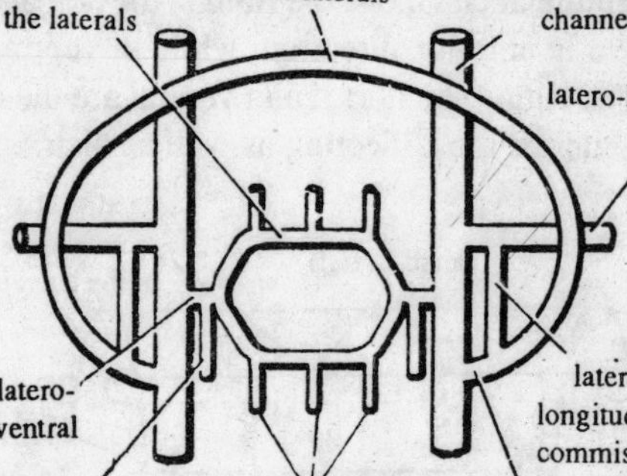

Fig 4.17 *Hirudinaria*. Diagrammatic representation of the lateral channels and their branches in the segment in dorsal view.

Excretory System.

The excretory system of leech comprises 17 air of segmentally arranged nephridia, situatea a pair in each of the segment 6th to 22nd. Of these the first 6 six pairs are located in segments 6th to 11th devoid of testis-sacs and are called the *pretesticular nephridia* and the remaining 11 pairs are the *testicular nephridia*. They remain in close association with the testis-sacs. The two types differ only in a minor respect.

Testicular nephridium. This is regarded to be a typical nehpridium. Its shape is just like that of a horse-shoe and consits of seven parts:

1. Main lobe
2. Vesicle duct
3. Vesicle
4. Apical lobe
5. Inner lobe
6. Initial lobe
7. Ciliated organ

1. *Main lobe*. The main lobe is actually just like a horse-shoe and is situated in the ventro-lateral position between the adjacent caeca

of the crop. In forms the major part of the nehpridium and consists of two unequal limbs, out of which anterior is long and the posterior limb is small.

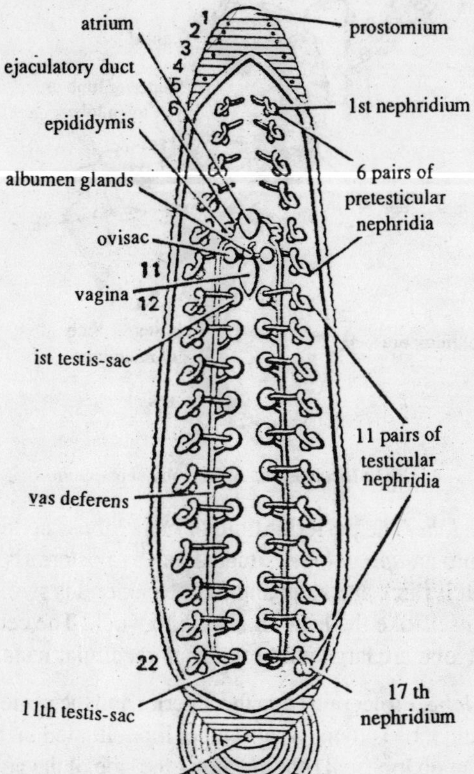

Fig 4.18 *Hirudinaria.* Showing reproductive and nephridia.

2. *Vesicle duct.* The vesicle duct is a narrow tube that starts from the inner end of anterior limb of the main lobs, runs backwards alongside the vesicle and opens into the vesicle near its hind-end.

3. *Vesicle.* The vesicle is a very large oral sac lying behind the rest of the nephridium and closely applied against the ventrolateral surface of the body wall, and opens to the exterior through a rounded aperture, the *nephridiopore.*

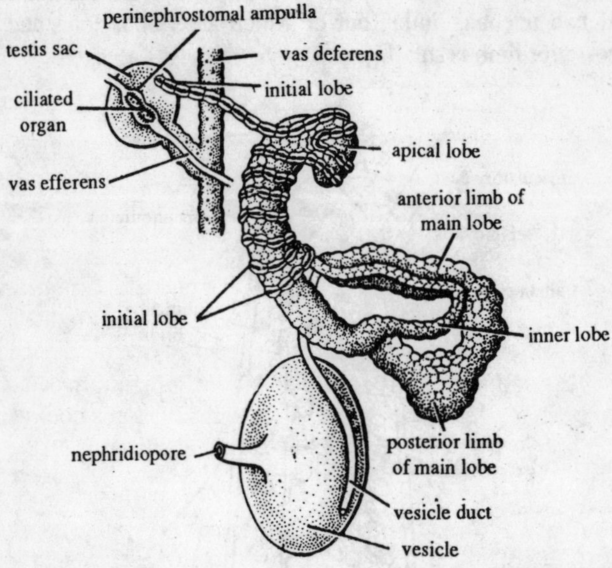

Fig 4.19 *Hirudinaria,* A testicular nephridium.

4. *Apical lobe*-The posterior limb of main lobe is continued anteriorly into an *apical lobe,* situated antero-posteriorly beneath the crop. It is thick and stout and its anterior end is swollen and bent upon itself like the head of a walking stick. The cells lining the apical lobe are large with regular intracellular canals.

5. *Inner lobe.* Projecting from the anterior and the posterior limbs of the main lobe is found a short inner lobe situated in the concavity of the main lobe and running along the side of the apical-lobe for about half its length.

6. *Initial lobe.* The *testis-lobe* or the *initial* lobe is a long, narrow, transparent and cord-like structure, formed of a single row of elongated tubular cells and closely twined around the apical lobe. Its posterior end joins the main lobe, while the anterior end runs inwards and reaches over the testis sac of its own side, where it ends blindly close to the *perinephrostomial ampullae.* The intracellular canal of the initial lobe gives off many diverticult in each cell.

7. *Ciliated organ.* The ciliated organ corresponds to the funnel or nephrostome of a typical annelid nephridium and each consists of

(i) a central reservoir, a *perforated chamber*, appearing more or less spongy, with an outer wall formed of a single-layer of cells and a central mass of connective tissue cells, the seat of manufacture of the coelomic corpuscles. (ii) *ciliated funnels.* numerous funnels, each fitting into a pore of the reservoir by its narrow proximal part, the *neck.* They cover the reservoir from outside Each funnel is like an ear-lobe, with about one-fourth of its margin incomplete along one side. The distal end is ciliated and broad.

In the adult *Hirudinaria* the ciliated organ has no connection whatever with the body of the nephridium although there is a distinct connection between the two in the embryo. In the adult it loses connection with the nephridium, gives up excretory function and associated with the haemo coelomic system, for which it manufactures coelomic corpuscles and drives them into the haemo-coelomic stream.

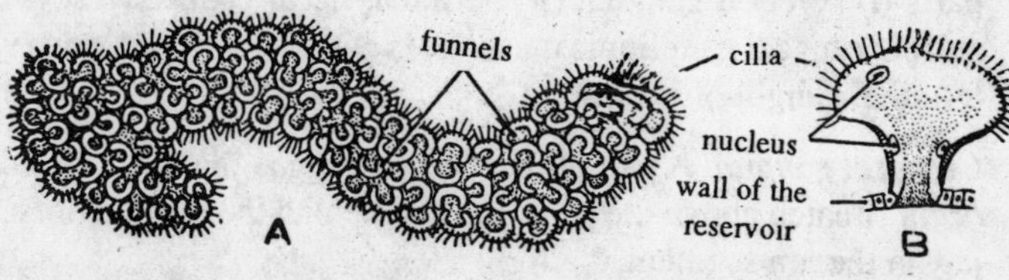

Fig 4.20 *Hirudinaria.* A= Ciliated Ordan., B = Single funnel.

Pretesticular Nephridia. The first six pairs of nephridia lying in segments from 6th to 11th have no relation with the testis sacs, since these are absent in these segments. Therefore, these are devoid of haemocoelomic ampullae and the ciliated organs. In other details these resemble the testicular nephridia.

Physiology of Excretion.

The nephridium proper is truloy excretory in function and serves to eliminate excess of water and nitrogenous wastes, (mostly ammonia and small quantities of urea). The nephridium is profusely supplied with branches of haemocoelomic channels. Its gland cells separate the waste products from the haemocoelomic fluid. The excretory fluid is finally collected into the terminal vesicle to be discharged to outside through the nephridiopore.

Several workers have also attributed an excretory function to the botryoidal tissue, the intracellular capillaries of which communicate with the haemocolelomic vessels.

Nervous System.

The nervous system satisfies the usual pattern of

Annelida. It consists of three parts.

1. Central nervous system,
2. Peripheral nervous system, and
3. Sympathetic nervous system.

1. *Central nervous system.* The entire central nervous system lies within the ventral haemocoelomic channel and consists of five parts-(i) cerebral ganglia, (ii) peripharyngeal connectives, (iii) sub pharyngeal ganglionic mass, (iv) ventral nerve cord, and (v) terminal gangolionic mass.

(i) *Cerebral ganglia.* A pair of fused celrebral ganglia form a small *brain* situated above the phariynx just behind the median dorsal jaw in the 5th segment.

(ii) *Peripharyngeal Connectives.* They start from the sides of the brain and pass downwards along the sides of pharynx to meet the supra pharyngeal ganglionic mass.

(iii) *Sub-pharyngeal ganglionic mass.* Ventrall the peripharyngeal connectives unite to form the subpharygeal ganglionic mass which lies below the pharynx exactly opposite to the cerebral ganglia. It is composite structure of more or less traingular shape, apex of the triangle pointing backwards formed by the fusion of four pairs of ganglia.

(iv) *Nerve-cord.* From the apex of subpharygeal ganglionic mass arises the *ventral nerve cord* It runs along the mid-ventral line from 6th to 26th segment. Although it appears single, it is really double as both are surrounded by a common sheath the *neurolemma.* The ventral nerve cord carries 21 well formed ganglia, each located in the first annulus of its own segment. Posteriorly,

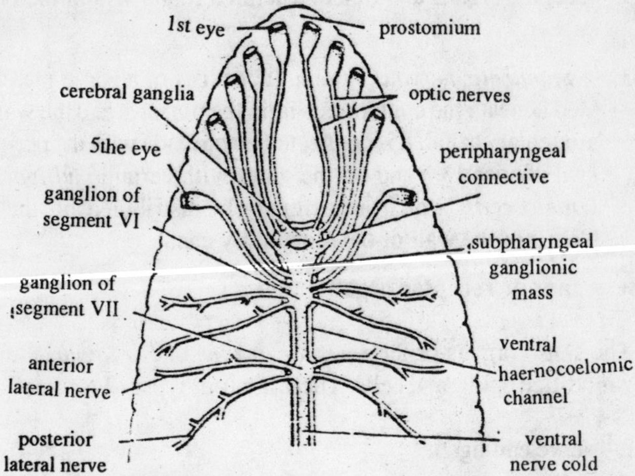

Fig 4.21 *Hirudinaria.* Nervous System. in anterior region of body

the ventral nerve cord ends in a large ovoid *terminal ganglionic mass,* situated within the posterior sucker. It is formed by the fusion of seven pairs of embryonic ganglia of the lasts even segments that constitute the posterior sucker.

2. *Peripheral nervous system.* It consists of several paired nerves which are originated from the central nervous system and innervate all parts of the body. A pair of stout nerves arise from the brain which run forward to supply the first pair of eyes, the prostomium and the roof of the buccal chamber. The subpharyngeal ganglionic mass gives out four pairs of nerves which go to the second, third, fourth and fifth pairs of eyes and also to segmental receptors of the first five segments, roof of the buccal cavity and the muscles of body wall. From each ganglion of ventral nerve cord arise two pairs of stout nerves, the anterior laterals and posterior laterals. The *anterior laterals* arise from the anterior part of the ganglion, branch and innervate the vas deferens, nephridium, the nephridial vesicle, muscles of body wall, ventral nerve receptors and two outer dorsal receptors of their own side. The posterior laterals arise from the posterior part of ganglion just behind the anterior laterals. The posterior lateral branch innervate the viscera, median dorsal region of the body wall and the central pair of the dorsal segmental receptors. The

terminal ganglionic mass sends off several nerves supplying the receptor organs and other structures found within the posterior sucker.

3. *Sympathetic nervous system.* It consists of a nerve-plexus, situated beneath the epidermis within the muscles and the wall of the alimentary canal. It is connected on one side with the peripharyngeal connectives and on the other with certain *multipolar ganglionic cells,* which are irregularly distributed on the nerve-plexus of the wall of the alimentary canal.

Sense organs or receptor organs.

The sense-organs or the *receptor organs* of *Hirudinaria* are especially modified epidermal cells. There are four types of receptor organs:

1. Nerve endings
2. Annular receptors
3. Segmental receptors
4. Eyes.

1. *Nerve endings.* These are the simplest and most numerous receptors. These are found between the epidermal cells all over the body with their ganglionic cells situated beneath the epidermis. These are simply the ends of nerve fibres. These organs are probably sensitive to chemical stimuli i.e. changes in the surrounding water. They, thus, act as *chaemoreceptors.*

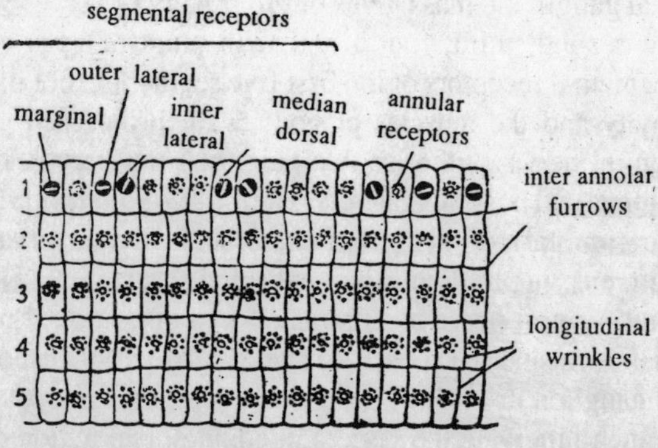

Fig 4.22 *Hirudinaria.* A segment showing annular and segmental receptors.

2. *Annular receptors*- These are in the form of minute papillae, formed of flattened overlapping epidermal cells. Their proximal ends are conneced with the nerve fibres while the distal ends project out of the integrument. About 36 annular receptors are found to be equally distributed on the dorsal and ventral sides in each annulus. These form transverse rows across the middle of the annulus. The annular receptors are tactile in function or *tangorecetors.*

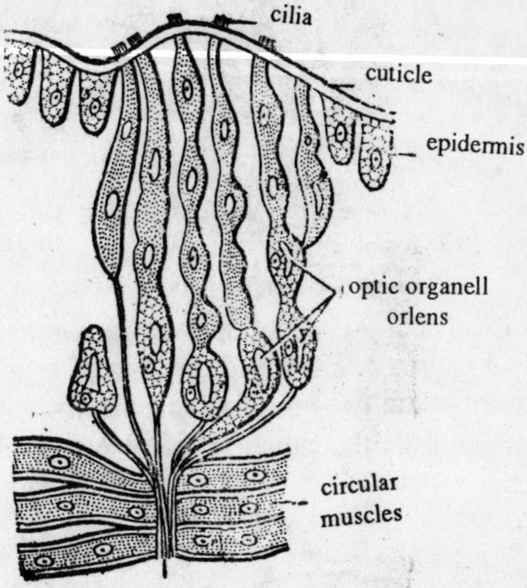

Fig 4.23 *Hirudinaria* An annular receptor in V.S.

3. *Segmental receptors.* These are small whitish elliptical papillae borne upon the first annulus of each body segment, four pairs dorsally and three pairs ventrally. Each receptor consists of two types of cells-tactile cells or *tangoreceptors* and light-sensitive cells or *photoreceptors.* There are 5 to 10 long, slender tactile cells, separated from one another and provided with hair-like processes at their outer free ends. The light-perceiving or the photoreceptor cells, found only in the dorsal receptors, contain a cresentic hyaline substance, the *optic organelle* or *lens,* in their cytoplasm. Each receptor receives a nerve branch and functions both as *tangorecepetor* and *photoreceptor.*

4. *Eyes.* There are five pairs of eyes lying in a semicircle of black spots on the dorsal surface of the anterior sucker, one pair in the first annulus of each of the first five segments. Each eye is

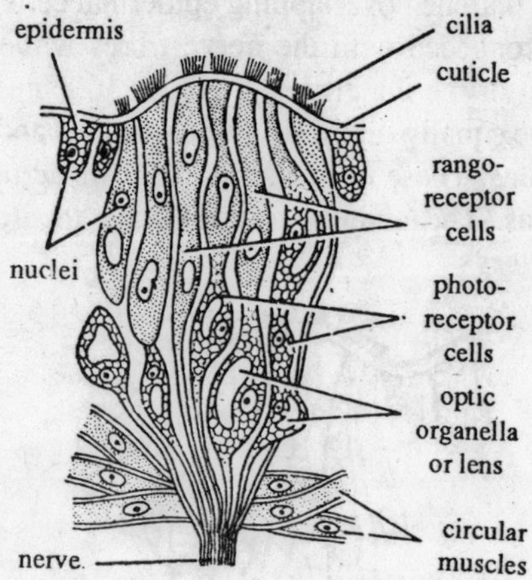

Fig 4.24 *Hirudinaria* A segmental receptor in V.S.

cylindrical or cup-like in shape with its long axis perpendicular to the body surface. Each eye consists of a long pigmented cup covered externally by transparent epidermis and cuticle forming a cornea. Inside the cup are refractive cells arranged in several

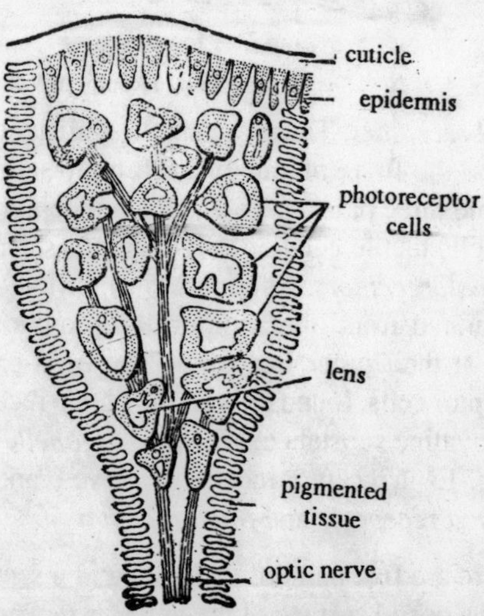

Fig 4.25 *Hirudinaria.* V.S. of eye

longitudinal rows. Each refractive cell contains a crescentic hyaline substance, the *optic organellae* or lens surrounded by a very thin peripheral layer of cytoplasm containing a small rounded nucleus. An optic nerve enters each eye basally and runs along its median axis distributing branches to all photoreceptor cells.

The eyes are directed in different directions and each can receive light from only one direction. It is not yet known whether images of objects are formed in the eyes. However, it is presumed that eyes can only distinguish light from darkness and can locate the direction of the source of light.

On the basis of their metameric arrangement and histological structure, *Whiteman* regards the eyes to be serially homologous with the segmental sense organs. Moreover, the segments with eyes lack the usual segmental receptors.

Reproductive System

Leeches are hermaphrodite *i.e.,* male and female reproductive organs are found in the same individual.

Male reproductive organs. male reproductive or generative organs consist of the following structures:

1. Testis-sac
2. Vasa deferentia
3. Vasa efferentia
4. Epididymis
5. Ejaculatory duct
6. Atrium.

1. *Testis sacs*-Testis sacs are eleven pairs of small, spherical or rounded vesicles found one pair in each of the segments from 12th to 22nd. These are located one on either side of the ventral nerve cord. From the wall of each testis sacspermatogonia are budded off which develop into spermatozoa.
2. *Vasa efferentia.* The spermatozoa pass from each testis sac into a short sinuous duct, the *vas efferens,* which arises from the

postero-lateralo border of the testis sac and runs outwards to join the *common vas deferens* of its side. All the vasa efferentia of one side open into the common vas deferens of that side.

3. *Vasa deferentia:* They are a pair of slender longitudinal ducts running forward, from twenty-second to the eleventh segment, lying along the ventral body wall on either side of nerve cord. Each vas deferens is enclosed within a tubular coelomic space which contains amoeboid corpuscles similar to those of the haemocoelomic fluid.

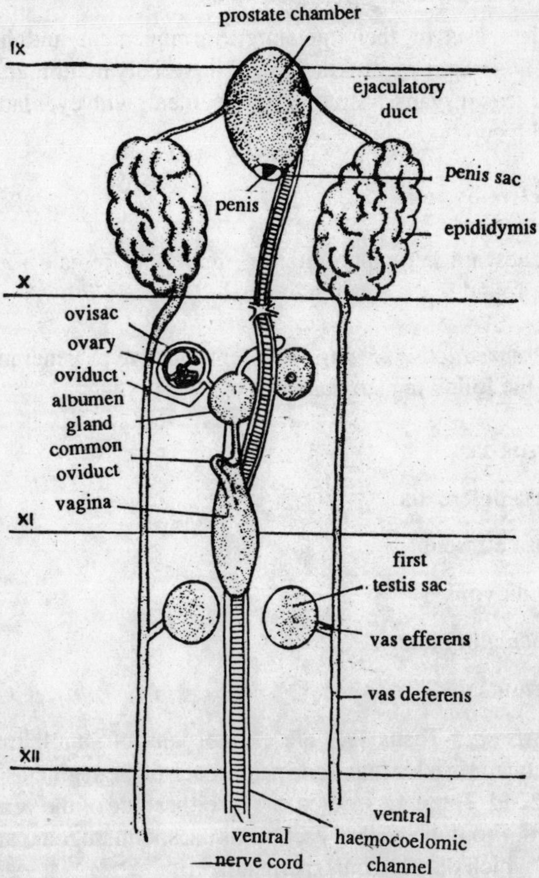

Fig 4.26 *Hirudinaria.* Reproductive System.

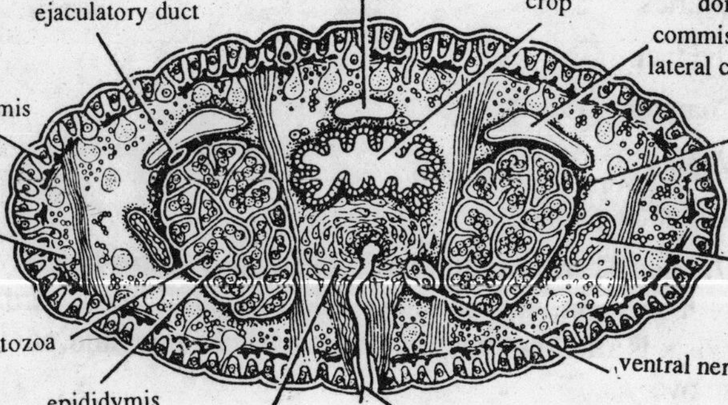

Fig 4.27 *Hirudinaria*. T.S. body through epididymis, prostate and penis sac.

4. *Epididymis.* The distal end of each vas deferens in the 10th segment widens into a coiled mass which is termed as *epididymis.* This is said to store sperm brought from the testis sacs.

5. *Ejaculatory duct.* Anteriorly, from the inner side of the epididymis arises a short narrow ejaculatory duct. It runs inward horizontally and joins the felolow of opposite side to form an *atrium.*

6. *Atrium.* The ejaculatory ducts of the two sides join a median pyriform sac, the *atrium,* extending into the 9th and 10th segments, and opening outside by the male genital pore. The atrium consists of two parts-a vase-like broad anterior part called the *prostate chamber* and a neck-like narrow posteriorly directed part, the *penis sac.* The *prostate chamber* has thick muscular walls which are covered externally by numerous unicellular *prostate glands.*

The penis-sac is the posteriorly directed narrow neck-like highly muscular part of the atrium which contains a coiled filmentous tubular *penis.* It helps in copulation and transference of spermatophores into the vagina of other leech.

Female reproductive organs. The female reproductive system includes:

1. Ovisacs
2. Ovaries
3. Oviduct.
4. Common Oviduct
5. Vagina
6. Albumen glands.

1. *Ovisacs.* There is a pair of ovisacs situated ventro-laterally in the 11th segment, one on either side of ventral nerve cord. Each ovisac is a hollow structure filled with coelomic fluid, containing an ovary.

2. *Ovaries.* A pair of ovary, each enclosed inside an ovisac. Each ovary is a delicate, coiled and filamentours structure which contains many nuclei. The ova are formed by the budding from the wall of ovary.

3. *Oviducts.* From the base of each ovisac a narrow oviduct arises which runs inward and backward. The right oviduct passes below the ventral nerve cord and joins with that of the left to form a common oviduct.

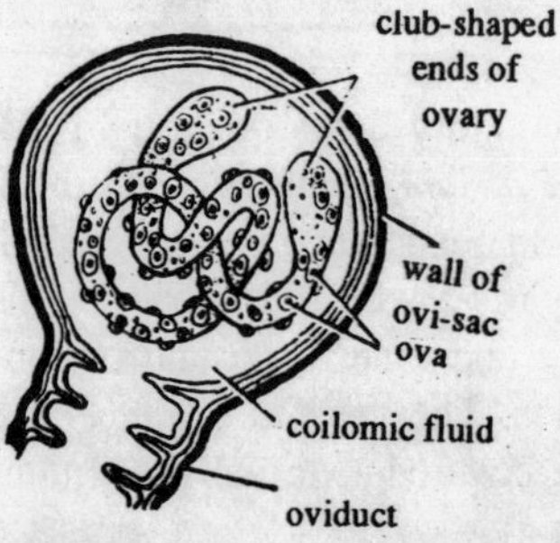

Fig 4.28. *Hirudinaria.* An ovisac showing ovary.

4. *Common oviduct.* The common oviduct is a mid-ventral type having its posterior past open folded and opening into vagina.

5. *Vagina.* It is a large and bear shaped structure, lying ventrally in the posterior past of the 11th segment. It enlarges considerably during the breeding season. It is continued anteriorly into a duct that opens out at the female genital pore. The female genital pore

lies on the ventral side in the groove between the second and third annuli of the segment 11.

6. *Albumen glands.* The place of the origin of common oviduct is covered by a thick layer of unicellular glands, the *albumen glands,* opening into the common oviduct.

The ova, budded off from the ovaries, pass through the oviducts into the vagina, where they remain till fertilization. The ova are coated with the albumen from albumen glands. Fertilization takes place in vagina. Fertilized ova are discharged into a cocoon for further development.

Copulation and fertilization

Copulation takes place in March and April. It may occur in water or on land and bests for about an hour. Two leeches come in contact by their ventral surfaces in head to tail position, so that the male genital pore of one lies against the female genital pore of the other. Each copulant protrudes its penis and passes it into the female genital pore of the other to shed sperms in the vaging. Thus cross-fertilization occurs.

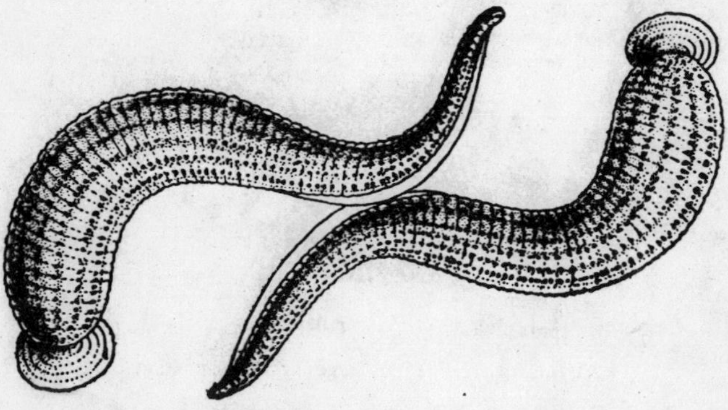

Fig 4.29 *Hirudinaria.* Two leeches inpulation.

Cocoon formation:

The cocoon is formed in the month of April to June after copulation. The cocoon is secreted in the form of snow-white, *frothy girdle* by the glandular skin of 9th, 10th and 11th segments. It is always formed during

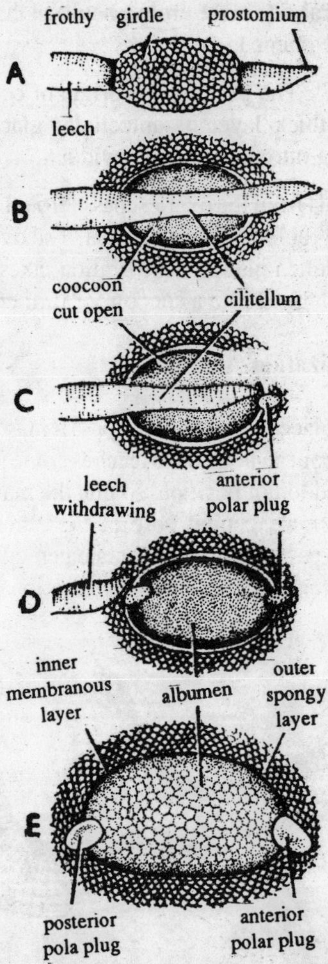

Fig 4.30. *Hirudinaria* Various stages cocoon formation.

breeding season. After the cocoon is formed, *clitellar glands* secrete an albuminous fluid, which is deposited in the cocoon with the fertilized ova. The cocoon is then passed over the head of each individual. Leech then withdraws its anterior end backwards by *rhythmic* movements. When the prostomial region of the worm comes in contact with the anterior end of the cocoon, prostomial glands secrete a kind of *plug* which seals the anterior part of the cocoon and is called *anterior polar plug*. The leech goes on withdrawing its anterior end backwards till the prostomial region comes in contact with the posterior end of cocoon. There, it also secretes the posterior plug. After the posterior plug is completed, the leech shakes off the cocoon, which is laid in a moist place by the side of a pond or pool but never in water. Cocoon becomes hardened on exposure to the air. Its colour is light-yellow or amber and it is barrel-shaped. It measures about 25 to 30 mm. in length and 12 to 15 mm. in diameter. The wall of cocoon consists of an inner, thin, tough membrane and an outer thick, spongy loayer. There are also *anterior* and *posterior polar plugs* on the anterior and posterior sides respectively.

Development

Development of fertilized egg takes place within the ootheca and takes about 15 days. In each cocoon 1-24 embryos develop. The development occurs at the cost of albumens. On completion of development,

5

ARCHIANNELIDA

The name *Archiannelid* was first of all suggested by*Hatsckek* in 1893. The group representing an assemblage of marine, elongated and thread like animals without set as or parapodia. Formerly they were considered as primitive forms, from whcih another annelids might be derived. The marine animals, included in this group, are Polygordius and Protodrilus. In additiokn to these main representative types different authors have included a number of other animals in this group i.e. Sedgwick included four other animals as: *Dinophilus, Saccocirrus, Aelosoma* and *Histriodrilus* Beriham in C.N.H. included only *Dinophilus* and exchuded all others. Cn the the other hand *Parker* and *Haswell* included *Dinophilus, Sacocirrus* and *Norilla.*

As representative type of Archiannelida, *Polygordius,* and Protodrilus may be taken as examples and the point of differences in which they differ from other forms can be delted later.

Characteristics of Archiannelida

Chief characteristics of Archiannelida as shown by *Polygpordius* and *Protodrilus* are as follow:

1. These are elongated, thread-like marine worms with cylindrical body.
2. Parapodia, setal, gills etc. are absent.

3 External segmentation indistinct but internal segmentation is complete and the segments are represented by coelom.

4. Small prostomium overhang the mouth and bears a pair of tentacles and a pair of ciliatxed pits. Prosatomium is followed by large peristomium or oral-segment which constitute the first segment of body.

5. Body is devoid of circular muscle fibres.

6. Nervous system is situated in epidermis and is not separated out from it. The ventral nerve cord lies in epidermis, is without ganglionic swellings. In *Protodrilus* two ventral nerve cords are widely separated from one another by the ventral groove, though in Polygordius the two ventral nerve cords are fused together along the length.

7. The right and left coelomic cavities are separated by dorsal and ventral septa. Oblique longitudinal septa are present which pass across the body cavity on each side, from the region of ventral nerve cord to the lateral body wall and in *Prododrilus* body cavity is traversed by space reticulum.

8. A pair of simple nephridia are present in each segment which open internally in segment preceeding in which the body of nepliridium lies.

9. Both dorsal and ventral blood versels are present in corresponding mesenteries but both are non-contractile. The colour of blood may be red, yellow, green or may be colurless depending upon the species.

10. Germinal cells are developed from coelomic epithelium and are discharged in body - cavity from where they pas out Detinite genital ducts are absent.

11. Sexes separate or united.

12. A typical trochophore larva, known as *Loven's* larva, is found in *Polygordius*, but it has not been reported from *Protodrilus*.

Development of Polygordius.

The larva of *Polygordius*, as it comes out from the body, is in the form of typical trochophore larva known as Loven's larva.

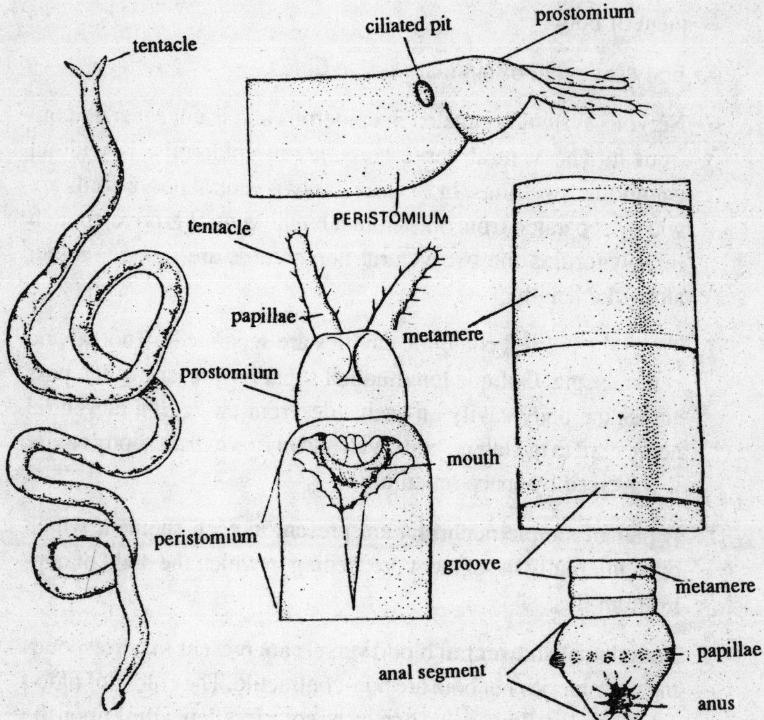

Fig. 5.1 *Polygordius*, external features. A-Entire worm; B-Side view of anteror end; C-Ventral view of anterior end; D-Segments of the trunk; and E - Anal segment

The larva is typically trochophore but is some what reduced in form. It has the appearence of a complete and wild cone united by the bases.

The larva bears a double row of cilia, the pre-oral cilia and post-oral cilia. The lateral girdle of cilia is somewhat weaker. Mouth lies at one side of larve between two ciliary bonds whereas the anus lies at the apex of post-oral cone and is surrounded by and girdle of cilia. The apical plats is present on the top of preoral cone and bears eye spot. There is a pair of meroblastic bands at the hinder nephridia. Alimentary canal is divided into stomodaeum, enteron and proctodaeum. The alimentary canal is separated from body wall by a vast space represented by blastocoel.

Metamorphosis. During metamorphosis posterior part of body elongates and, mesoderm becomes separated to form somites, forming trunk region of the adult. This elongated region grows due to active growth of post-oral cone and gets segmented both, internally as well as externally. The anterior part of larva perish in adult though it greatly reduced and forms the head of adult. The base of apical plate forms the brain and thickening of epidermis on ventral side is termed as ventral nerve cord. As development proceeds the mesoblastic somites become rolled out to form mesoblastic cavities.

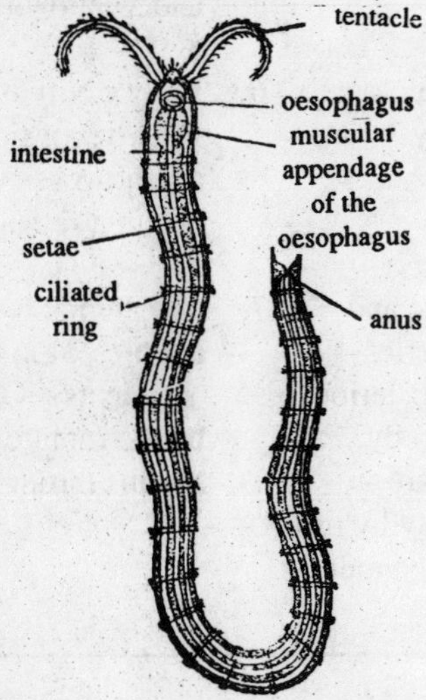

Fig. 5.2 *Protodilus*

Differences between *Polygordius* and *Protodrilus*.

Polygordium	*Protodrilus*
1. Found in sand of European Seas.	1. It occurs in sand of European inlands, sea-lake at Faro near Messina.
2. External segmentation is shown by faint grooves.	2. External segmentatioon is marked by ciliated grooves.

3. A median ciliated groove is absent.	3. A median ciliatedgroove is present.
4. Ventral nerve cords are fused along mid-ventral line.	4. The ventral nerve cords are widely separated from one another being situated on either side of mid-ventral groove.
5. A short invisible buccal region is present followed by oesophagus.	5. Buccal region is not visible but there is a 'U' shaped muscular tube placed ventral to oesophagus and opens into it, which is eversible help the animal in burrowing and swallowing mud.
6. Coelom brightly coloured and devoid of any reticulum.	6. Body cavity is lined by a network of fibres which are, however, few in number forming reticulum.
7. Sexes are separate and the testes and overies are confined to posterior segments of the body. Ova and sperms are escaped from body by its rupture, because gonoducts are absent.	7. The animal is bisexual. Testes are present in the posterior segments. Ova are released by the rupture of body wall while sperms through the nephridia.

Dinophilus.

Dinophilus is included in class Archiannelida by several workers. Its structure is much similar to those of *Polygordius* and *Protodrilus*. It is a small worm found in English Channel on sea weeds during spring. It consists of a head constituted by a pre-oral and post-oral chambers followed by a trunk which is, in an immature animal, distinct to show 5 or 6 segments. Each segment is demarked externally by one or two ciliary rings. Pre-oral lobe of prostomium is broad and bears a pair of simple eyes, one or two ciliated pits. Tentacles are absent.

The alimentary canal is straight and ciliated. There is a ventral muscular protrusible organ opens into eosophagus. The nerve cord is continuous with epidermis and consists of a brain or ganglion in prostomium and a pair widely separated nerve cords. The two ventral nerve cords sometimes show incipent segmental ganglian.

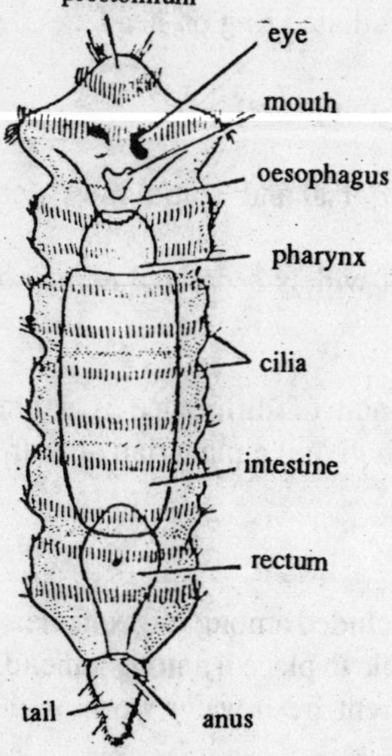

Fig 5.3 *Dinophilus*.

Body is traversed by a neat work of connective tissue in which sex-calls are present. Five pairs of protonephridia opening on the surface of the body. Internally no inter--segmental septa or dorsal or ventral mesenteries or ducts.

Sexes are separate, the male being much smaller than female. The part of nephridia in males are connected with testes and act as seminal vesicle and the male genital products are passed through the common aperture formed by the union of 5 pairs of nephridia before they open out. A median penis is present which passes through the body wall. The young ones are hatched as trochophore larva.

Saccocirrus

It resembles *Protodrilus* in many respects so tthat it must be placed close to this animal. Following characters show resemblence with *Protodrilus:*

1. Presence of ciliated ring on head.

2. Presence of oblique septa.

3. Presence of dorsal and ventral mesenteries.

4. Presaence of widely separated nerve cord which is in contact with epidermis.

The only point of difference is a presence of segmentally arranged setae, which give it a planarian appearence.

Aelosoma

It is often included among oligochaetes, but from its characteristics it seems desirable to place it among annelids of uncertain position. It is a small, transparent fresh water worm with a loose and uncertain body.

1. The prostomium is ciliated ventrally and ciliated pits are present.

2. Four setae are present in each segment behind prostomium.

3. Epidermis contains coloured cells, circular and longitudinal muscles are present but both are very thin.

4. One pair of nephridia present in each segment.

5. Central nervous system possesses a cerebral ganglion or brain which is continuous with epidermis but not followed by ventral nerve cord in trunk region.

6. Reproduction is by fission. Gonads are present but gonoducts are absent. The male gametes or sperms pass through a pore on the ventral

side of 6th segment. In addition spermathecae are present in segment 3-5.

7. Bundles of setae are present on body wall.

Nerilla

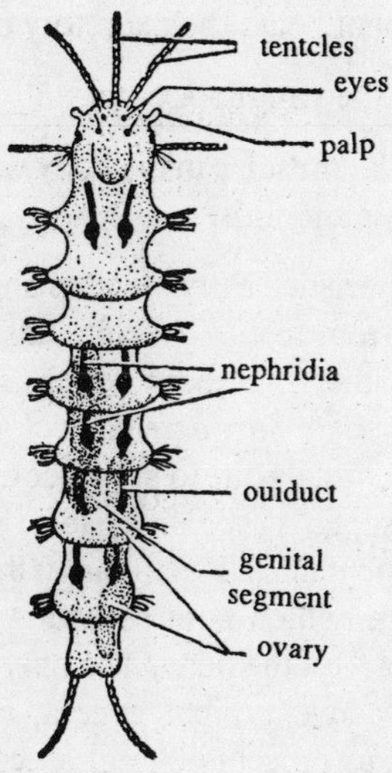

Fig. 5.4 *Nerilla*

Nerilla is included among Archiannelida by *Parkar* and *Haswell*. These are minute marine worms with segmentally arranged parapodia. Their position, however,is extremely doubtful whether it should be annelid or archiannelid.

Affinities of Archiannelida.

The main animals which are included in this class are *Polygordius and Protodrilus*. These are definitely annelids as is shown by the fact that they have segmented body, intersegmental septa and segmentally arranged nephridia.

But they may be regarded as primitive members of the phylum because of :

1. Absence of setae or parapodia.
2. Primitive condition of nerve cord which is in contact with epidermis.
3. Even the ist segment bears the excretory organ.
4. In the nerve cord the ganglia are not fully formed.
5. The layer of circular muscles in the body wall is not developed as is present in higher annelids.

Due to above characteristics *Polygordium* and *Protodrilus* must be placed in separate class from rest of the annelids i.e. in class Archiannelida. As regards the other members which are included by some authors in class Archiannelida, *Saccocircus* is closely related and allied to the above mentiokned animals due to presence of the segmented body, coelom, septa, oblique septa, vascular system, segmentally arranged nephridia and prostomial tentacles. It also shows the primitive characters of central nervous system, which is in contact with the epidermis and presence of the ciliated pits on the head. The only point of difference in which it differs from the two typical genera, *Protodrilus* and *Polygordius,* is that *Saccocirrus* presents lateral caecae in each segment in which it resembles higher animals, but even in this character it is very close to *Protodrilus.* As regards *Dinophilus* it is included in Archiannelida due to presence of a body which is extraordinary segmented specially in young stage. But it shows much structures than even that of *Protodrilus* and *Polygordius.* It differs from them in following respects:

1. Absence of tentacles.
2. Body without coelomic or oblique septa.
3. Locomotion by ciliary bands.
4. Absence of vascular system.
5. Body is traverced with network of reticulum.

Thus on the whole it seems advisible to place it in separate sub-class. In Archiannelida the typical forms are like *Polygordius* and *Protodrilus.*

6

METAMERISM

Division of the animal body into an anterio-posterior row of similar parts, or seqments, is called *Metamerism.* Each metamere typically repeats some or all of the various organ units. The term metamerism is applied only when organs of mesodermal origin are so arranged. Each segment or metamere contains nephridia, blood vessels, reproductiveor-gans, nerves, muscles etc. The segments work in co-operation with all other segments. They are integrated into a single functional unit. The segmental organs are interdependent.

Metamerism is always limited to the trunk region of the body. The head, represented by the prostomium and beasing the the brain and sense organs and the pygidium represented by the terminal past of the body which carries anus, are not metamers. New segments arise just infront of the pygidium, thus the oldest segments lie just behind the head.

Occurrence of Meta merism

The metameric repetition of body organs in first seen in Annelida in the animal kingdom. It is also of considerable importance as the most evolved group of animal kingdom i.e. Arthropoda and Vertebrata also show their body parts metamerically repeated. In annelids, the metamer-ism is cleasly visible both externally as well as internally, but in asthropods, it is mainly external. In man and other vertels rates there is an internal segmentation of body muscles, nerves, vestebral and few blood vessels.

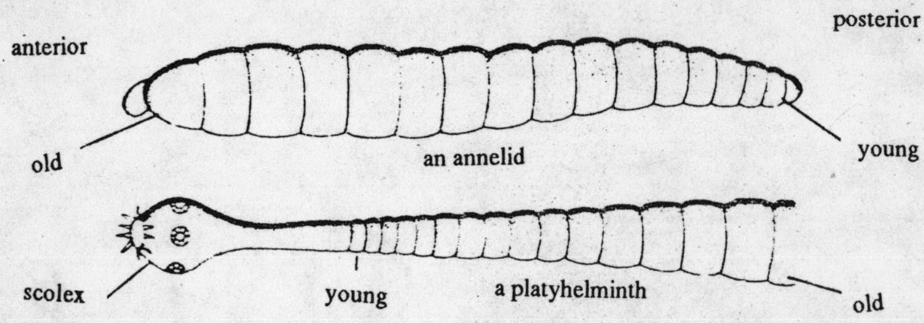

Fig. 6.1 True metamerism and pseudometamerism.

Thus metamerism is a feature of evolutionary advancement that has contributed towards the greater complexity of animal bodies.

Types of metamerism

The metamerism is animals can be of following types:

1. True metamerism
2. External metamerism
3. External and metamerism
4. Complete metamerism
5. Incomplete metamerism
6. Homonomous segmentation
7. Heteronomous segmentation
8. Psendometamerism

1. *True metamerism.* True metamerism is seen in annelids, arthropods and chordates. In true segmentation, the segmentation of the body is based on the segmentation of mesoderm. New segments are formed posteriorly.

2. *External metamerism.* In arthropod the metamerism is external, internally the segments are not market by partitions.

3. *External and internal metamerism.* Metamerism is well market in most annelids. This metamerism is external as well as internal.

Internally, the segments are separiated by transverse partitions or septa. Only the alimentary canal remains uneffected.

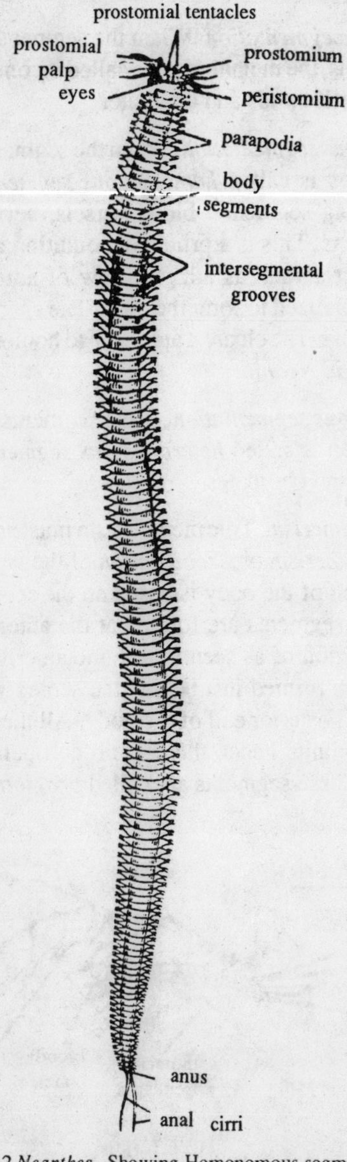

Fig 6.2 *Neanthes*. Showing Homonomous segmentation

4. *Complete metamerism.* When segmentation is seen in all the organs, the metamerism is called complete metamerism. Example: annelids.

5. *Incomplete segmentation.* When the segmentation is not seen in all the organs, the metamerism is called incomplete metamersim. Example: arthropods and chordates.

6. *Homonomous segmentation.* When the segments are all alike, the segmentation is called *homonomous segmentation.* Each metamere having segmental blood versels, nerves, nephridia and coelomoducts. This is a primitive condition and is not found in any existing animal as atleast a few of anterior segments are always specialized to form the head. The specialization is called *Cephalization.* The cloarest approach to homonomous segmentation is seen in *Neanthes.*

7. *Heteronomous segmentation.* If the segments are dissimilar, the segmentation is called *heteronomous segmentation.* Examples: arthropods and chordates.

8. *Pseudometamerism.* True metamerism must not be confused with *pseudometamerism* or *strobilization* of the tapeworms. Here the segmentation of the body is based on the segmentation of ectoderm. New segments are formed at the anterior end unlike the posterior region of as seen in true metamerisms. The youngest segments are formed just behind the scolex where as oldest lie towards the posterior end of the body. All the segments work as indepentent units hence there is no co-operation between the segments. These segments are called *proglottids.*

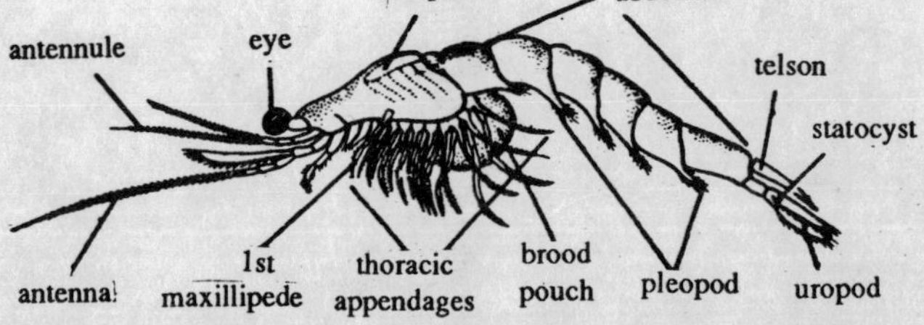

Fig. 6.3 *Mysis*

Origin and evolution of metamerism.

There are conflicting views regarding the origin of metamerism. Various hypotheses have been proposed to explain the origin of metamerism, but all of them are lacking to provide ample evidences and thus none is acceptable. The various hypotheses are as follows:

1. Corn or fission theory
2. Psendometamerism theory
3. Cyclomerism theory
4. Locomotory metamerism theory
5. Embryological theory

1. *Corn or fission theory.* According to this theory, a segmented animal is really a chain of individual zooids. Each segment represents a zooid. The supporters of this theory believe that metameric segmentation resulted when some non-regmented ancestor divided by transverse fission repeatedly or by asexual buddings producing a chain of sub individuals or zooids, united end to end due to their incomplete separation. These segments gradually become integrated morphologically as well as physiologically and merged into one complex individual.

The Chief objection to this theory is the lack of gradations of age in

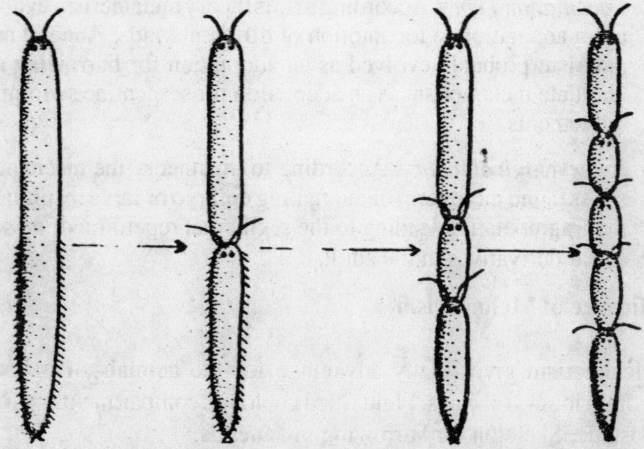

6.4 Origin of metamerism by fisssion

such a chain of zooids, which is true of segments in a metameric animal, such as an annelid. Another objection is that reproduction by fission in usually confined to the fixed animal, where as the ancestors were probably free-swimming.

2. *Psudometamerism Theory.* According to another view, in the ancestral forms, the various body organs had repeatedly spread out in the entire body. Hence the metamerism had resultled by the segmental division of these organs, such as nerves, nephridia, blood vessels and coelom. Within a single individual and not by the formation of new individuals as stated by fission theory. The division of body into segments occured as a result of the formation of cross-partitions or septa with each segment receiving a repeated past of each segment. This theory is supported by *Hyman* (1951) and *Goodrich.*
3. *Ceyclomerism Theory.* This theory was proposed by *Sedgwick* (1884) and supported by *Remane* (1950, 1963). This theory assumes that coelom originated in some ancestral actinozoan coelenterate, through the separation of four enteroocoelic pouches from the gut or paragastric cavity. Further division of two pouches into three pairs of coelomic cavities viz. protocoel, mesocoel and metacoel in the ancestral coelenterate. Loss of protocoel and mesocoel led to the formation of unsegemented coelomates such as molluses. Later sub division of metacoel produced primary segments leading to the segmented annelids.
4. *Locomotion Thory.* According to this theory metamerism evolved as an adaptation to locomotion of different kinds. Annelid metamerism probably evolved as an adoptation for burrowing and chordate metamerism as an adeptation for sespentine swimming movements.
5. *Embryological theory.* According to this theory the mechanical stress in the mesoderm of elongating embryo or larva resulting in the fragmentation leading to the segmental repetition of mesodermal derivatives in the adult.

Significance of Metamerism

Metamerism gives many advantage for the animals. It helps in locomotion in several ways. Fluid filled coelomic compartments provide hydroostatic. Skeleton for burrowing in annelids.

Metamerism offers *division of labour.*

7

COELOM IN ANNELIDA

The coelom is a perivisceral cavity between the body wall and alimentary canal. It is lined by peritoneal layers, on the outer side by a parietal layer of mesoderm and on the inner side by a visceral layer of mesoderm. It is filled with a coelomic fluid dsecreted by peritoneal layers. It protects the inner organs from external shocks. It is derived from the mesoblastic somites of embryo which swells up and surround the intestine primitively, therefore, the body cavity is divided into a series of paised cavities. Two types of septa are formed, one separating the cavaties of someties adjacent to each other, these are known as *transverse septa.* The to her is the one separating the cavities of two somites of a pair on dorsal and ventral sides of aliment any canal. These are known as *dorsal* and *ventral mesenteries.* The mesenteries break down in the adult but persists in Archiannelida.

Organs in association to coelom.

These are two organs associated with coelom. They are the excretory organs or nephridia and generative organs. The nephridia are the portion of the coelom and retain a communication with it by the nephrostomes or ciliated funnels. The generative organs are always products of the coelomic walls having thickened patches of coelomic epithelium.

Coelom in Polychaeta.

Coelom is well developed in Polychaeta and generally divided by the transverse septa into chamber, except in tubicolous polychaetes.

Fan-like oblique muscles pass from the mid-ventral line, starting from the lateral sides of nerve cord. They extend obliguely dorsally and outwards spread out in a fan-live manner and ends in the dorso-lateral body wall. These muscles divide the body cavity into these regious-two lateral and a mediam which hasbours the alimentary canal. These muscles are present in the middle region only of each segment at the label of parapodia. Burrowing polychaetes have no transverse septa and in *Glycera* they are absent in the anterior region to accommodate the eversible proboscis. Septa are absent in first 40-50 segments of Terebellides except between 4-5 segments. This is thick and is known as diaphragm.

Coelom in Obligochaeta

Coelom is divided by transverse septa into chambers which communicates usually above the nerve cord at the place where the nerve cord enters the next segment. In *Lumbricus* these are perorations in septa and are guarded by sphincters to regulate the flow of coelomic fluid. The coelomic fluid is colourless with amoeboid corpuscles and communicates with the exterior through three sets of apestures:

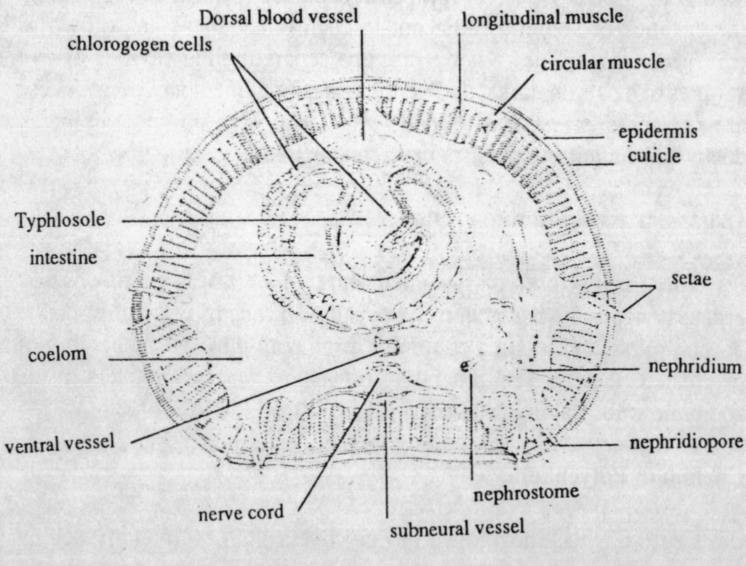

Fig. 7.1 *Lumbrious* T.S. through middle region of the body.

(i) nephridia (ii) genital ducts and (iii) dorsal poses. The Coelom is lined by flat epithelium except some pasts of alimentary canal which are lined with choragogen cells. Dorsal and ventral mesenteries are absent in adult. However, in *Branchiura* mesenteries are present even in adults.

Coelom in Hirudinea.

The presence of a true coelom has been established by the study of its development and its relation to the renal organs. It is always more or less obliterated by connective tissue into a system of communicating spaces and is comparatively present in *Acanthobella. Johanson, Oka, Seger and Shatia* have worked on true nature of the channels in the body of leeches. They are coelomic in nature but secondarily acquired the muscular walls. There are no true blood-vessels in *Hirudo* but their place is taken up by Coelomic channels which are filled with blood-like red fluid, *Shipley* has shown connection between the coelomic channels and blood vessels. The haemocoelomic system in Hirudinea is a peculiar communication of body-cavity and blood vascular system.

Body cavity. In *Acanthobella*, the coelom is specious and is divided into chambers by septa. In others it is broken up but retains a relation with some of thee organs such as ventral nerve cord, nephridia and main blood vessels. It is largely without a perivisceral nature but has the form of longihidinal canals connected by complicated system.

Blood vascular System. Blood is red with haemoglobin in Gnathobdellida and colourless in Rhynchobdellida. Since *Leydig* the blood vessels are distinguished from coelomic channels by the thicker and muscular walls of blood vessels.

Haemocoelomic System. The sinus system is after held to be a past of the vascular system through the reparation of main channels. These is a communication in the finer branches of the two systems, but according to *Oka* and *Seger* such connection is not visible.

In *Acanthobella* coelom forms a link between oligochaeta and Hirudinea. There is a spacious coelomic, incompletely divided by segmental septa. The coelom is enchroached upon by the growth of parenchymatus tissue and reduction of body-cavity.

In *Glossiphonia* the sinus system consists of a median dorsal sinus above the alimentary canal, a median ventral sinus on ventral side, two lateral sinuses and two intermediate sinuses consisting of a network of canals connecting the median and lateral sinuses on each side. These is also a system of hypodermal sinuses beneath the skin communicating with lateral and intermediate sinuses. Dorsal sinus encloses ventral blood vessel, ventral nerve cord and openings of nephridia.

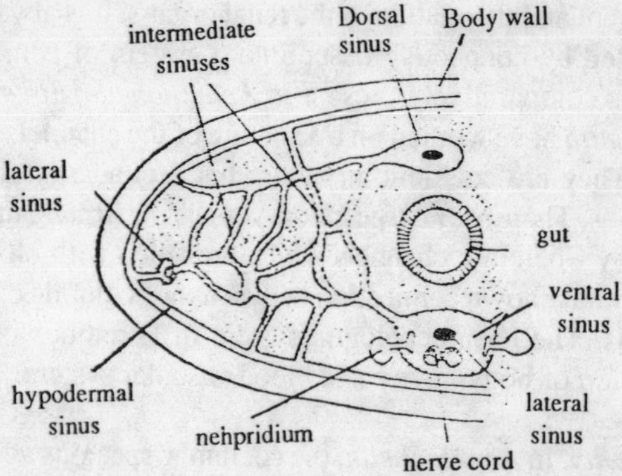

Fig. 7.2 *Glossiphonia*. T.S.

The fluid contains two types of Corpuscles: the smaller one like blood corpuscles and larger ones are from epithelial lining of sinuses.

In *Pontobdella* these are four main sinuses: a dorsal, a ventral and two lateral ones. The dorsal and ventral sinuses enclosing the corresponding blood vessels and so do the lateral ones. The ventral sinus also contains the ventral nerve cord and the ovaries. the branches of ventrala sinus surround the testes and nephrostomes. The lateral sinuses are contractile having muscular wall.

In Gnathobdellida, the coelom is greatly obliterated. It is caused by a tissue known as *botryoidal* tissue which possesses channels by their inter-communication, the coelom and vascular systems are brought to contact. Botryoidal tissue is brown in colour and embraces the alimentary canal, blood vessels and main lacunae of coelom. It consists of swollen cells joined together in anastomosing rows and containing tubular cavities. These contains a red fluid and constitute a tubular network between alimentary canal and body wall.

Haemo coelomic System in Hirudinaria. Two principle blood-vessels are lateral in position uniting with each other at anterior and posterior ends of the body. They give out branches ventrally and dorsally which anastomose with one another. The smallest branches break up in fine remifications in the integument, nephridia etc. The dorsal sinus lies just above the alimentary canal. The ventral nerve cord. These two sinuses are in communication with each other posteriorly and also by means of their branches with capillaries of skin, setting an indirect connection between blood vessels and coelomic sinuses. The sisnus which the nephridial funnel opens are in communication with ventral sinus.

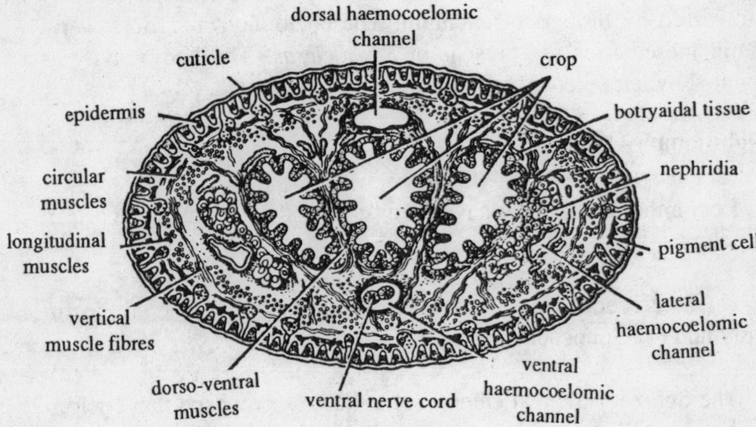

Fig. 7.3 T.S. of *Hirudinaria* through crop.

In *Nephelis* these are no dorsal and ventral blood vessels but only two lateral vessels are present. Dorsal sinus is stotally absent but ventral and two lateral sinuses are present. Lateral sinuses receive the nephridial funnels.

Relation between sinus system and botryoidal vessels. The botryoidal vessels communicate with dorsal and ventral sinuses or with their extensions. From their development these can be no doubt of their coelomic nature. They contain a blood similar in colour and characteristics of blood. This has led to view that the two systems are continuous with each other. But according to *Sedgwick* it is extremely unlike that animals so closely similar in other respects as Rhynchobdellida and Gnathobdellida should differ in such a important point of structure. Secondly sinuses and small blood vessels are intermingled in great complexity and it is difficult to distinguish between finer branches one be mistaken for other. Lastely the continuity between the vascular system and sinus system should be a unique phenomenon in the animal kingdom. However, it is at present well established that these do exist relation between these two systems, at least through finar remifications.

Coelom in Archiannelida.

The coelom is spacious in Polygordius with dorsal and ventral mesenteries and oblique septa across the body. In *Protodrillus* the body is traversed by the space reticulum. The coelomic septa, dorsal and ventral mesenteries are present in *Saccocirrus*. The body cavity is occupied by lateral connective tissue in which genital cells appear.

Evolutionary Origin .

Four important theories have been suggested regarding the of origin of coelom:

(1) Enterocoelic theory (2) Schizocoelic theory (3) Nephrocoelic theory and (4) Gonocoelic theory.

The Schizocoelic and enterocoelic theories proposoe that coelom originates by splitting of the mesoderm in which case the cavity is called a *schizocoel* or by evagination of pouches from the wall of the archenteron in which case it is called entersocoel.

The enterso coelic theory was suggested by *Sedgwick* (1884) with a specific aspect of Coelenterate organization that coelom might have been derived from coelenteric pouches. *Clark* points out that it is not easy to explain the selective advantages of such a step. The schizocoelic theory although seems at first sight plausible yet it has received little support. It does not explain the clear-cut morphological differences between the coelom and the spaces of the blood vascular system.

Nephsocoel theory was suggested by *Lankester* in 1874, at that time when the relationship of Coelom with nephridia and coelomoducts was not cleasified. Hence this theory has only historical interest.

8

FEEDING IN POLYCHAETES

Feeding is one of the stages in nutrition in which the animals collect and capture the food from the medium. The polychaetes have exploited a wide range of feeding nitches. They live in different ways. Some polychaetes are crawling, some are pelagic, some are tubicolous, a few are burrowing and others parasitic. Based on the mode of feeding the polychaetes are classified into the following types:

1. Carnivores or raptorial feeders
2. Browsers
3. Sand and mud feeders
4. Filter feeders

1. *Carnivores or raptorial feeders.* It includes most crawling, some tubicolous and all pelagic forms. They feed on small crustaceans, sponges, molluscs etc.

The raptorial feeders have adaptations for capturing their prey. The parapodia are well adapted for active locomotion as it is essential for running after the prey. The food is captured by means of an eversible pharynx or proboscis as in. *Neanthes.* The mouth is ventral and surrounded by the peristomium. Mouth opens into the phasynx which is lined with cuticle. The pharynx is armed with a single pair of chitinous jaws. Du ring feeding, the buccal cavity can be everted and the pharynx can be protruded out. The protruded pharynx is called proboscis. Eversion is rapid placing the jaws at the anterior end of the body and causing them to open. The prey is caught by the jaws and the pharynx is retracted. The protrusion of proboscis is brought about by the working of a set of muscles called *protractor muscles,* the reversion is brought about by the *refractor muscles.*

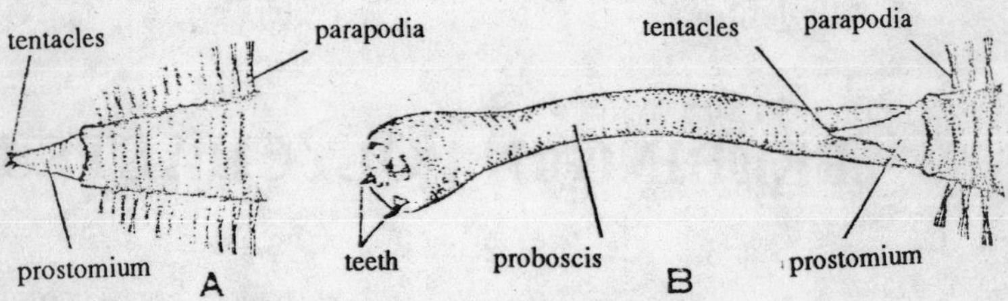

Fig. 8.1. Glycera. A = anterior and B = Head with protruded prboscis.

The number of teeth varies in the different species. *Eunice* has a pair of lower jaws and about five pairs of upper jaws. *Autolytus* has a circlet of teeth. In *Glycera* at the tip of proboscis four teeth are present.

The length of proboscis also varies in the different polychaetes. in *Glycera,* the long tubular proboscis can be protruded to about 1/5 the length of the body, when retracted it occupies about the first twenty body segments which lack septa. The proboscis of *Autolytus* is long and rectractile lying within the buccal cavity.

2. *Browsers.* Browsers are herbivorous polychaetes feeding on algae. They use their jaws to tear off small pieces of algae. Example is some jaw bearing errant polychaetes, particularly some nereids.

3. *Sand and mud feeders.* The sand and mud of sea bottom contains organic matters such as bacteria, diatoms and other micro-organisms. This deposit matter has great nutritional value and is an important source of food for many invertebrates. Certain polychaetes depend on these matters and are called sand and mud feeders. These polychaetes are of two types:

(i) Direct deposit feeders

(ii) Indirect deposit feeders

(iii) *Direct deposit feeders.* These are burrowing or tubicolours animals. They swallow the mud and sand as such. In the alimentary canal, the microorganisms are digested and absorbed.

The undigested food, mud and sand are eliminaled in the form of castings. Direct deposit feeders include a number of burrowing polychaetes such as *Arenicola, Scalibergma, Ophellids, Capitellids* etc. As the mud and sand contains only a small amount of food material, the animal should swallow a large amount of sand. The muscular pharynx helps swallowing.

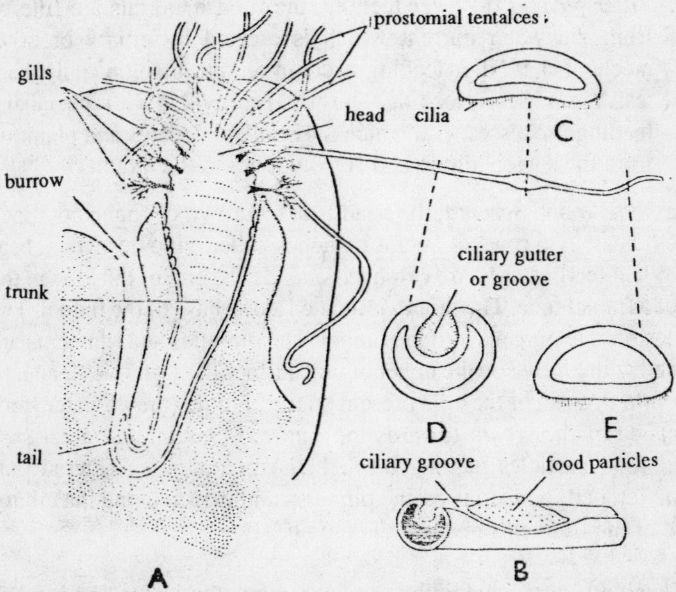

Fig. 8.2 Terebella. Tentacle structure and feeding. A- Worm in feedinng posture. B-A section of tentacle rolled up to form a ciliary gutter. C- C.S. of tentacle creeping over substratum. D- C.S. of tentacle creeping over substratum. D- C.S. of tenntacle showinng ciliary groove. E - Tenntacle turned over and creeping.

(ii) *Indirect deposit feeders.* It is also called *tentacle feeders.* The proboscis is absent here, instead, they possess ciliated tentacles which are extended over or into the substratum. *Terebella* lives in a tube. The head bears numerous long, hollow muscular tentacles. The tentacles can be moved in a variety of ways. One side of the tentacle from base to the tip is ciliated. This side of the tentacle can be folded to form a ciliated groove along with the food is propolled into the mouth. Surface detritus adheres to the mucous secreted by the tentacular epithelium. The food particles are moved down the groove by ciliary act ion while large particles are forced along by a requeezing action of the sides of the groove.

Similar mode of feeding is seen in *Amphitrite* several other groups of polychaetes have evolved tentacle-feeding independently eg. cirratulids, spinoides, *Magelona* etc.

4. *Filter feeders.* In filter feeding, the food materials are filtered from the water current which is created by ciliary or other mechanisms. Filter feeding is common in tubicolous polychaetes. A proboscis is lacking. The head is provided with specialized feeding processes with which they collect detritus and plankton from the surroundidng and convey them to the mouth.

In *Sabella* and *Serpula*, the head bears a crown of filaments. Each filament bears two rows of lateral branches called pinnules which bear cilia. When feeding they are extended outside the end of the tube in the form of a funnel tube. The mouth is located at the base of the funnel. The pinnules secrete mucous into the funnel. The pinnules and filaments are provided with groove on the upper or oral surface. The grooves lead into the mouth. Beating of the cilia present on the pinnules draws water from the outside of the crown towards the centre. Plankton and other suspended food particles passing through the pinnules are filtered out, become trapped in mucous on the pinnules and driven along the ciliary groove of the filaments towards the mouth.

Chaetopterus is quite unique and specialized in its feeding mechanism. It is a tubicolous polychaete lives in a 'U' shaped parchment tube. The body is differentiated into anterior, middle and posterior region. The middle region is formed of five segments and it is modified for filter feeding. In the first segment of the middle region, a pair of wing like structure, the *great wings.* The notopodia of the second segment are modified into a cuplike structure, the *food cup*. It is present on dorsal side and is ciliated. A *food groove* runs forwards and opens into the mouth. It arises from food cup. The notopodia of the next three segments are fused together to form three *fans*. The extremely long and aliform or wing-like

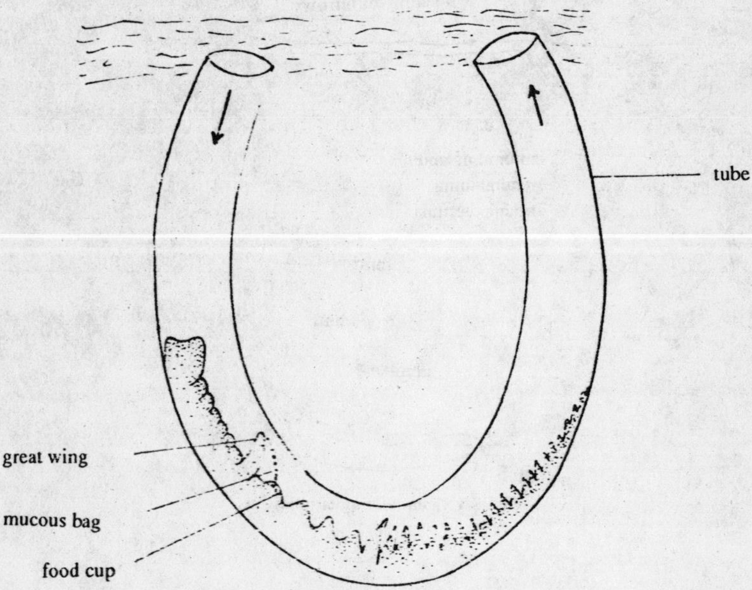

Fig. 8.3 *Chaetopterus* in its tube

notopodia of 12th segment arch over the back, their tips touching to form a ring. Into this ring mucous is recreted by the greater wings. The fans move to and fro and creat a water current. The water entering the tube must pass through the mucous which strains out all the suspended matter. As this mucous bag is loaded, it is carried into the food cup where it is rolled into a ball. The ball moves along the food groove to the mouth. A new mucous bag is secreted every 20 minutes and about 290 ml water passes through each bag.

Arenicola marinus lives in a 'J' shaped tube. The head lies towards the small limb of the tube. The worm ingests sand continuously towards the blind-horizontal end of the burrow, by means of a simple short eversible proboscis. As more and more sand is swallowed, a funnel like depression is formed on the surface. Water enters the tube from the depression, percolating down through sand which filters suspended planktons. This organically rich sand is swallowed by the worm.

Periodically the worm stops feeding and backs up the vertical limb of the burrow and defaecate in the form of castings.

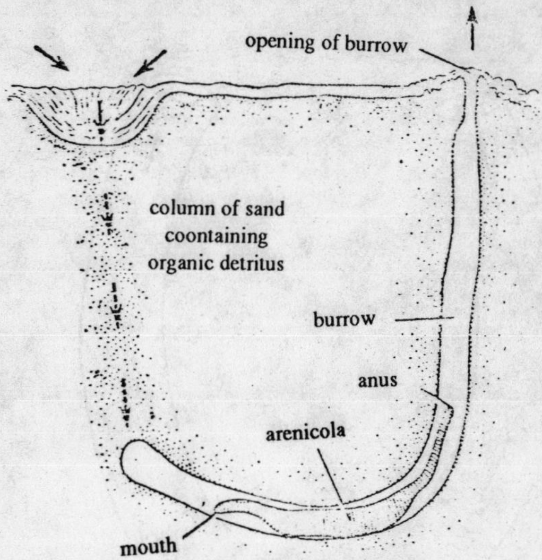

Fig. 8.4. *Arenicola* in its tube.

9

LIGHT EMISSION IN POLYCHAETA

Production of light without heat is one of the most striking feature of the living organism and excite ones' interest. This phenomenon is known as *bioluminescence*. Phylum Annelida includes many living lumineous forms out of which most are from Polychaeta and few from Oligochaeta. However, no such forms are known to occur among Hirudinea and Archiannelida. The discovery of lumineous polychaete can be probably attributed to *Auzout* and *De La Voice* (1666) who observed worms on oyster. Later on several other workers have added to our knowledge and the modern work on luminescence has begin with that of *Panceri* (1878) who investigated many species of polychaetes belonging to five families.

1. Family Aphroditidae.

The luminescence is restricted to the scales app ear to a segment which cover the dorsal surface in two rows. *Panceri* (1898) noted that the luminious region might cover the entire region of scale except for the point of attachment or a half-moon shaped area of light depending upon the species. The light appears only on stimulation and it may be extend from head to tail or tail to head. *Haswell* (1882) has shown that when stimulus is powerful flashes of phosphorescence may run along whole series of elytra. It seems that phosphorescence has a protective action. Several view have been expressed regarding the source of light in these animals.

I. Initial view was that of *Quadrefages* (1893) who wrongly thought that muscles were the source of light.

II. The second view came from *Panceri* who distinguished nerve endings in *Polynoe* as the source of light because of abundence of nerve fibres in the elytra. *Jorden* (1885) suggested, on the basis of histological studies, that the secondary cells of hypodermal layer on the under-susface of scale functions as photogenice cells.

Curiously *Kutsch* (1909) who has extensively worked came to the different conclusison. He found in *Acholoe* species that at the base of the papillar gland, cells arranged like a star which are suppose to secrete a luminous material into the sea water through a *Caudal pore*.

The latest studies on the scales of luminous worms is that of *Bonhomme* (1940-41) who believed that Kutsch's photogenic cells at the pore of papillae are ganglionic cells and papillae are sense organs. He could not find any external secretion on the scales and he believed that the light was intracellular.

Physiology. Polynoe emits light only on various kinds of stimulations such as thermal, chemical, mechanical or electrical. There are certain controverscy as regards the direction in which they commenses and passes. *Falgeis* (1918) in his experiment showed that process of light production undoubtedly is by oxidation.

2. Family Alcipidae

Greef (1885) described an Alcipidae to which he attributed the ability to produce light. *Okada* (1805) discovered a luminous sponge but it was actually due to the annelids that filled in the canal system of sponge.

3. Family Tomopteridae.

A series of yellow structures, visible to the naked eyes are present on the parapodia of these colourless worms. Some designated them as eyes while others thought them as glands. Although the details were studied by *Hor ovey* (1926) and *Mayer* (1930). There are two types of light organs large or small on the parapodia. Each organ is formed of a ring of yellow presumably photogenic cells forming a sort of sheath. A nerve ganglion is present on the inner end of the light organs with nerve fibres, running between the transport cells to the base of the yellow cells. Luminescence is presumably intracellular and light is emitted on stimulation.

4. Family Syllidae

The members of this family present a more striking luminous display than all other worms. The worms have got peculiar feature of swarming for breeding purpose. Several workers have made interesting contribution to such an effect. The luminescence of *Odontosyllis* is undoubtly a mating adaptation serving to bring sexes together at the time when the eggs and sperms are ripe.

Histology. *Galloway* and *Welch* found two kinds of cells in epidermis of *Odontosyllis*. The first type is rather large and the second type is shaped varrieties with the appearence of spiral twist. Each cell has distinct neck and open to exterior through a pore in cuticle. Since the later were empty, they were thought to be phosphorescence organ but *Dahalgren* (1916) was inclined to doubt that the twisted cells were photogenic and regarded them as mucous cells.

Horway have been able to demonstrated the presence of luciferin and luciferir in two species of *Odontosyllis*.

5. Family Nereidae

Knowledge of most luminous nereids is in a most unsatisfactory state as it is difficult to identify the worms that were designated as *Nereis* by earlier workers. *Forskal* (1762) and *Viviani* (1805 were probably syllids. There appears to be no observation of any nereids producing light. *Hardman* (1903) desceribed heteronerids in gulf of Mannor but only suggested that the light come from wreids and the account is convincing.

6. Family Chaetopteridae.

Panceri first described the detailŝ of luminescence in *Chaetopterus*. *Panceri* has shown that there are five organs which produce luminescence in *Chaetopterus*. These are:

(i) Base of feeller or anteenae.

(ii) Gland cells of base and of dorsal side of wing like process or pinnules of the middle body region.

(iii) Suctorial disc which has the form of a smalol sac.

(iv) Edges and upper surface of three paddles of middle region.

(v) All the notopodia of posterior region.

However, the regions of greatest luminescence do vary in different species. It is also possible that entire *Chaetopterus* is luminous but the observation is not yet sufficient to emphasize that statement with certainty.

Histology. The epithelium which is responsible for luminescence is a complex structure different in different regions because it is also connected to the formation and repair of parchment tube which the animal never leaves.

Panceri (1878) has designated photogenic cell of two kinds :

(a) Spherical cells present as special masses forming the gland of pinnula.

(b) Peryform cells scattered along the ciliated epithelium of luminous region.

Dalf abserved two types of cells supposed to be photogenic :

(i) Small, oval ones or flask-shaped with ducts.

(ii) Elongated type extending across the epithelium

Almost similar observations were made by several workers. *Von Horni* (1943) who made careful histological studies using stains with an affinity for mucous. *Fuziwara* (1935) studied the histology in *Mesochaetopterus japonicum* where he has shown the three kinds of gland cells :

(A) Club-shaped Cells with neck and pore Scattered throughout the pithelium and containing rather small granules.

(B) Closly packed elongated cells along the ridge of the tentacles and in other luminous region filled polycrome blue+ve granules.

(C) Oval Cells below the surface of the epithelium containing fibriller contents and very small eosinophillous granules.

It is ceertain that elongated columnar cells containing tolidin blue or polycrome blue are photoogenic cells but the photogen to mucigen is not clear and must await for researches on the biochemistry of light production.

7. Family Cirratulidae.

A few species belonging to this family has been reported to bioluminescent. Intensity of light increasees in summer months. The luminous secretion comes out from paired lateral region in the posterior part of the worm. The integument shows mucous cells whose number and size increases in summer period. The large mucous cells are considered undoubtly to be photogenic.

8. Family Terebellidae.

A number of workers has described the light of Terebellides polychaetes emit light from both, the long head-tentacle and body while *Polyccirrus* emits light through body. The muscle of the tentacles and unicellular glands with a secretory pore and scattered region only in the epithelium. *Dilgron* (1910) was able to distinguished clearly the cells containing large masses of yellow material which gave yellow colour to the tenacle. He has shown eight types of cells some of them are representing stages of development. Some cells contain granules and other rods. The former type is believed to be bioluminescence.

10

EXCRETION IN ANNELIDA

Excretion is the process by which the metabolic waste products are thrown out of the body. In annelids the process is performed by special organs called nephridia (*Lankaster,* 1877). These nephridia exist a variety of modification, not only in different species, but also in the different segments of the body of same individual. In addition to nephridia the chloragogan cells, coelomoducts etc. may play accessory role in excretion.

Excretion in Polychaeta.

There are two types of ducts which communicate the coelom to exterior. These are

1. Nephridia
2. Coelomoducts.

A variety of combination exists between the two above and this aspect has seen studied by *Goodrich* (1940), who also traced the evolutionary status.

Types of Nephridia. There are two types of nephridia :

(i) *Protonephridia.* It is the primitive type of nephridium. It opens to the exterior by an opening called *nephridispore.* But internally, it terminates in the coelom as a blind tube. Internally the tube branches. The closed ends of branches bear a number of *solenocytes* which help in excretion. The solenocyte resemble the flame cell

It has a cavity surrounded by a mass of cytoplasm with a nucleus in it. The lumen of these cells enclose a long, vibratile flagellum (in each). Excretory fluid enters through the walls of the nephridial tubules which are internally ciliated. The fluid is driven into the lumen of the nephridium by the flagellum and forced to the exterior through the nephridiopore. The protonephridia are found in *Glycera,Phyllodoce, Nephthys* etc.

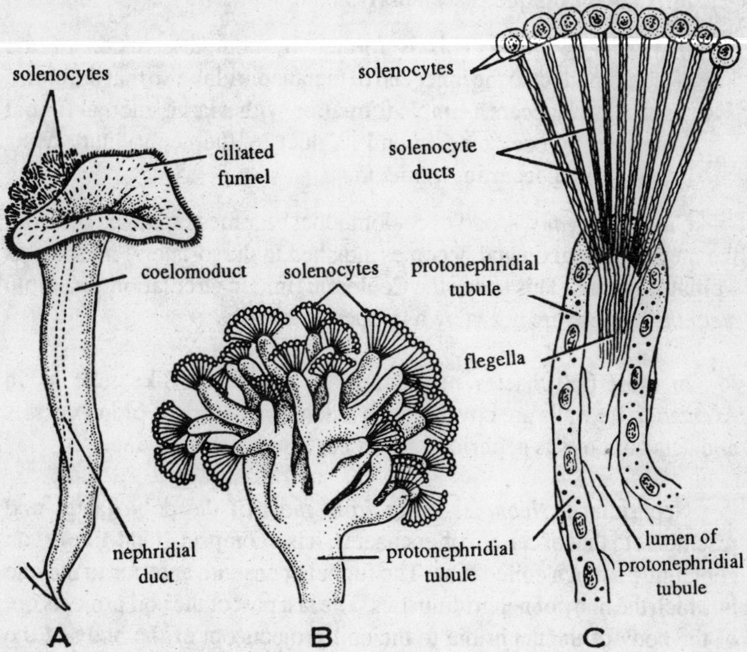

Fig. 10.1 Protonphridium and coelomoduct in *Phyllodoce paretti*.
A-Relation of protonephridium and coelomoduct. B-Branched end of protonephridium. C-Solenocytes of one protonephridial branch.

(ii) *Metanephridia*. These nephridia have two openings. One end of the nephriddium opens into coelom by a funnel like opening, the *nephrostome* and the other end opens out by nephridisspore *Coelomoducts*. These are segmentally arranged, wide tubes developed as out pushing from the coelom to the exterior. The Coelom communicates to the exterior with the help of coelomoducts. They open into the coelom by a large ciliated funnel called *coelomostome*. Basede on the combination occuring between nephridia and coelomoducts, *Goodrich* (1945) recognized following types of structures called *nephromixia* :

(a) *Protonephromixium* This is formed by the coelomoduct becoming grafted on to the excretory canal of a protonephridium. These are seen in *Phyllodoce.*

(b) *Metanephromixium.* This is formed by the coelomoduct becoming grafted on to the open nephridiostome of a metanephridium. A good example is seen in *Hesione.*

(c) *Mixonephridium.* This is formed by complete fusion of the coelomoduct to the inner end of metanephridium so that the entire structure appears a simple formation with a large internal funnel formed by coelomoduct and its duct by the nephridium. Mixonephridia occur in *Arenicola.*

Ciliated Organs. A part of coelomoduct becomes separated off from the metanephromixia and becomes attached to dorso-lateral muscles as a ciliated organs. This keeps the Coelomic fluid in circulation. Example *Neanthes.* These are not known to open externally.

In most polychaetes nephridium is a wide sac-like tube as in *Arenicola.* Its walls are covered with a dense net-work of blood vessels and acts not only as nephridial organ but also a gonadial duct.

Nephridia of *Neanthes* differ from most of the polychaetes and resembles to that of certain oligochaetes. It is a compact gland-like organ containing a much coiled tube. The funnel opens into anterior to the one in which the body of nephridium lies. The last past of the coil projects out of the body of the nephridia to the coil projects out of the body of the nephridia to the exterior at the base of parapodium.

Location and number. Generally the nephridia are present in almost all the segments of the body with the excepetion of a few on either side. But in *Arenicola*, there are only six pairs and in *sternastis* only one pair occur. In tubicolous polychaetes the anterior nephridia are enlarged and the posterior are reduced. In the fan-worm, a single nephridial canal opens on the head.

The principle nitrogenous waste in polychaetes is ammonia. The waste diffuses from the coelomic fluid or the blood to the nephridial tubules and then transfered to the lumen to be removed to out side.

Excretion in Oligochaeta

Except *Paranis uncinata* all the obigochaetes have nephridia. The nephridia which are present are of metanephridial type possessing an internal pre-septal funnel or nephrostome, its tubules and forms a terminal nephridiopore.

With our present growing knowledge several corrections have been made in the classification of oligochaetes nephridia. *Stefenson* had recognized only eight types while *Bhal* classified them into thirteen types. *Bhal* has also discovered the enteronephric.

Bhals' classification is obviously far superior to those of earlier workers, since he took into consideration not only their size but also their structure, development and phylogenetic relationships.

The large nephridia are called *meganephridia* where as small are called as *micronephridia*. These are small as compared to meganephridia and are not related to septa. There is one another type known as *mesonephridia*. The mesonephridia can further be divided into *exonephric* (opening out-side the body) and *enteronephric* (opening into the alimentary canal.

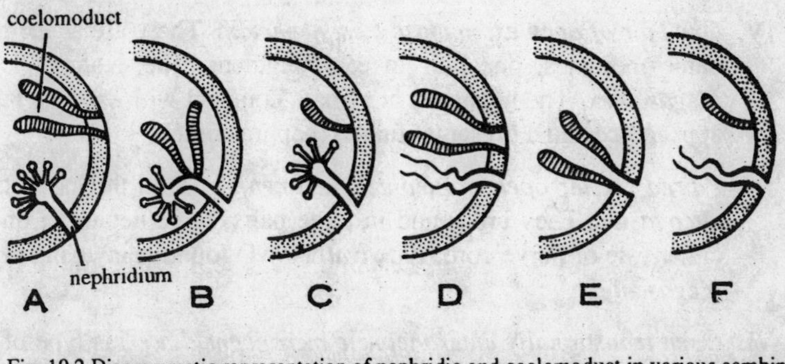

Fig. 10.2 Diagrammatic representation of nephridia and coelomoduct in various combinations in Polychaeta. A-Protonephridium and separate coelomoduct (*Vanadis*). B-*Protonephromixium* (Phyllodoce). C-Protonephridium and ciliated organs (Nephthyidae and Glyceridae). D-Metanephridium and separate coelomoduct (Capitellidae). E-Mixonephridium (*Arenicola*). F-Metanephridium and ciliated organ (Neries).

Holonephridia. These are generally large in size and did not undergo any division during their development. The holonephridia can be divided into following types

I. *Open exonephric holonephridia.* This type is found in *Lumbricus trestis.* The nephridia is divided into pre-septal funnel (nephrostome) and post-septal body. The body is divided into

(i) a long narow tube

(ii) these is a short brownish tube which is ciliated

(iii) a wide past of tube and

(iv) muscular tube and reservoir.

II. *Open transtionally exonephric holonephridia.* These are found in *Allololesphora* species from 35 segments onwards but never in 1-34 segments. The nephridia unite to two longitudinal excretory canals running parallel to the alimentary canal. These canals unite at the opening of anus with the alimentary canal forming a median ampulla.

III. *Closed exonephric holonephridia.* These are found in *Chaetogaster.* These nephridia are 'U' shaped and without cilia and nephrostome.

Meronephridiai. The nephridia breaks down or splits up during the course of development and are generally small in size. They can be classified into following types:

IV. *One pair of open exonephric mesonephridia.* These are as their name indicates, one pair in each segment. The example is *Megascolex.* The nephridia are small, compact with sac-like in appearence and a preseptal funnel (nephrostome).

V. *Paired cluster open exonephric meronephridia.* In the case of *Dicogaster.* They are found in three pairs, three nephridia on either side of nerve cord. *Vidyavati* (1941) found many pairs in *Megascolelx.*

VI. *Open transtionally enteronephric meronephridia.* This type of nephridia are seen in *Heplochaetella* from 3 segments onwards. The nephridia on either side joined by a pair of longihidinal excretory ducts as in transitional holonephridia, but the ampulla is totally absent. The Canals open near the anus in alimentary canal.

VII. *Open enteronephric mega-mero-nephridia.* Due to large size of the nephridia attain, *Bhal* has considered them as mega-mero-nephridia. Since earlier workers were misled by their size. There a free septal funnel or nephrostome. Body is composed of straight and twisted lobes. The canal opens into supra-intestinal excretory canal which runs abovoe the alimentary canal and opens into alimentary canal. Examples: *Pheretima* and *Megascolex*.

VIII. *Closed exonephric meronephridia.* These are found in 5-9 segments of *Lampito*. Each nephridium has got its own tubule which opens into pharynx *Tufted nephridia.* These nephridia originated either due to the incomplete branching of holonephridia or union of the terminal canals or ductules of mesonephridia. They are classified into following types.

X. *Open exonephric tufted nephridia.* They occur in the pre-clitellar and clitellar segments of *Pontoscolex* species. The number of nephridia varies being 80-100 in second segment and they gradu-allydecrease in number.

XI. *Closed exonephric tufted nephridia.* These are 9-10 pairs of tuft of nephridia occurs in each segments. Example is *Megascolex.*

XII. *Closed enteronephric tufted nephridia.* They occur in 4, 5 and 6 segments of *Pheretima* and are often described as pharyngeal nephridia.

Chloragogen cells. In earthworm the intestine is surrounded by some yellow cells called *chloragogen cells.*

They take off nitrogenous waste products from the blood capillaries of the gut and deposit them as yellow granules called *guanin* in their cytoplasm. When fully loaoded with waste products, they detach themselves from the gut wall and drop into the coelomic fluid. In the fluid they are consumed by the amoebocytes. The debris of these cells are wafted towards the nephrostome and are taken out by nephridial tubes. According to some workers chloragogan cells serve as store house for reserve materials.

Samal Van Gansen (1956) mentions that in terrestrial species of oligochacta, the silicates got from food material and soil are removed from the body and deposited in the chloragogen cells as waste products. Liebman (1946) describes that the eleocytes which are free cells into the coelom formed as a result of release of chlorgogen cells are laden with food reserve. However, Semal-Van Gansen regards eleocytes as degenerate chloragogen cells which are destroyed by phagocytic coelomcytes.

Excretion in Hirudinea

In hirudinea the excretory organs are segmentally arranged nephridia. These are 17 pairs lying in segments 6-22. The anterior 6 pairs of nephridia are situated in segment 6-11 which are without testis sacs, hence called the *pre-testicular nephridia.* The posterior 11 pairs of nephridia are called *testicular* nephridia as they occur in segments 12-22, each of which contains a pair of testfis-sacs.

Each nephridium is a horse-shoe shaped structure called *main lobe.* The main lobe has two limbs-an *anterior-limb* and a *posterior limb.* A lobe-like structure lies between the two called inner lobe. The posterior limb is continued forwards as the *apical lobe.* The anterior end of the apical lobe looks like the handle of walking stick.

There is a long hollow tube which coils around the apical lobe called *initial lobe.* One end of the initial lobe opens into the main lobe and the other ends blindly in the testis sac. From the inner end of the anterior limb arises a duct called *veside duct.* The vesicle duct leads into a *vesicle.* An *excretoory duct* arises from the vesicle and opens out by an opening called *nephridiopore.*

Ciliated organ. Inside the perinephrostomial ampullae lies. The *ciliated organ.* A ciliated organ has a central *reservoir* which is perforated, around which are innumerable *ciliated funnels.* The reservoir is a more or less spongy structure with an outer wall formed of a single layer of cells and a central mass consisting of connective tissue cells and corpuscles. Each ciliated funnel is like an ear-lobe, with about one-fourth of its margin incomplete along one side and is covered with outwardly directed cilia on its outer margin and inner surface. The ciliated organ manufacture coelomic amoeboid corpuscles. Ciliated organs are joined to nephridia in the embryo, but in the adult animal they have no connection with nephridia. This showos that originally the ciliated organ belonged to a nephridium, but in the adult, having no connection with the nephridium. It is not excretory but has become a part of haemocoelomic system for making corpuscles. Ciliated organs correspond to coelomoducts of Polychaeta and not to nephrostomes of nephridia.

11

ASEXUAL REPRODUCTION IN POLYCHAETA

Reproduction is the process by which organisms give rise to new individual. It is a characteristic feature of all living creatures for maintaining their race. Mainly there are two types of reproduction (i) Asexual and (ii) sexual. In the asexual reproduction the daughter individuals are developed without the involvement of gametes. These daughters are developed exclusively from the somatic cells of parents.

The power of reproduction is closely associated with the power of reproducing lost part and with the power of definite growth i.e. the growth after the adult condition has been attained. In many polychaets the number of segments continue to increase throught the life of animal by the addition of new segments between penultimate and anal segments and in most, if not, of them the power of regeneration is very extensively. Asexual reproduction takes the form of budding or fragmentation, processes that involve both growth and regeneration of lost part.

Asexual reproduction in polychaetes have been extensively studied by *Malaquin* (1893), *Potts* (1913) and *Okada* (1934). Polychaetes exhibits three types of asexual reproduction. These are:

1. Fragmentation
2. Transverse fission
3. Budding

1. Fragmentation

In the fragmentation the body breaks into small pieces and each piece or segment develops into a complete individual. The fragmentation is used in natural production of new individuals in two ways by the polychaetes.

(a) *Spontaneous fragmentation.* The division of body into single segments or short sections occurs spontaneously. This is followed by regeneration of a complete in *Chaetopterus* any segment isolated from the anterior region develops into a complete and whole animal. This ability is retained only by the anterior fourteen segments only. The anterior and posterior ends of isolated segments grow in length until the normal number of segments have been produced anteriorly and posteriorly. Other examples are *Ctenodrilus, Phyllochaetopterus, Autolytus* etc.

(b) *Orderly fragmentation.* It occurs in *Sabella Pygospio* etc. The posterior part of body breaks off by a transverse fission to form a new individual.

2. Transverse fission.

In the transverse fission,l the parent divides into two pieces and each piece develops into a complete individual. A constriction develops at a definite region called *zone of fission.* Epidermis grows into the zone of fission to form a partition called *macroseptum.* This is the point where constriction develops. The anterior piece regenerates the anal region and the posterior region develops the head. The transverse fission takes place in following animals.

(a) In *Salmacina* and *Sabella* a constriction appears towards the posterior end. separating a few segments. Tentacles grow out just behind the constriction so that the posterior part becomes a distinct zooid. The latter finally breaks off and develops the full number of segments. The parent regenerates the missing segments. The parent regenerates the missing segments.

(b) In *Chaetogaster diaphanus.* additional segments are proliferated at the hind-end of the worm. A constriction develops infront of the last 5th segment. Fresh proliferation occur in front of the segments further forwards. This results in a chain of zooids that remain jointed together for sometimes ultimately, the zooids get detached and develop sex-cells.

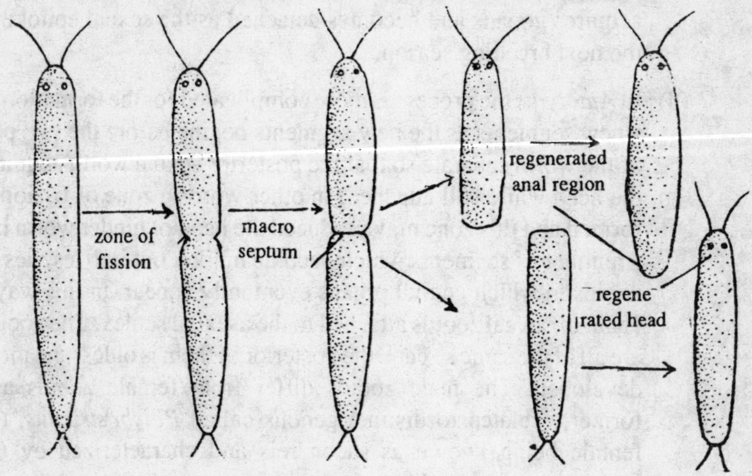

Fig. 11.1 Processs of transverse fission

(c) In *Odontosyllis*, the body has two regions anterior *atoke* and

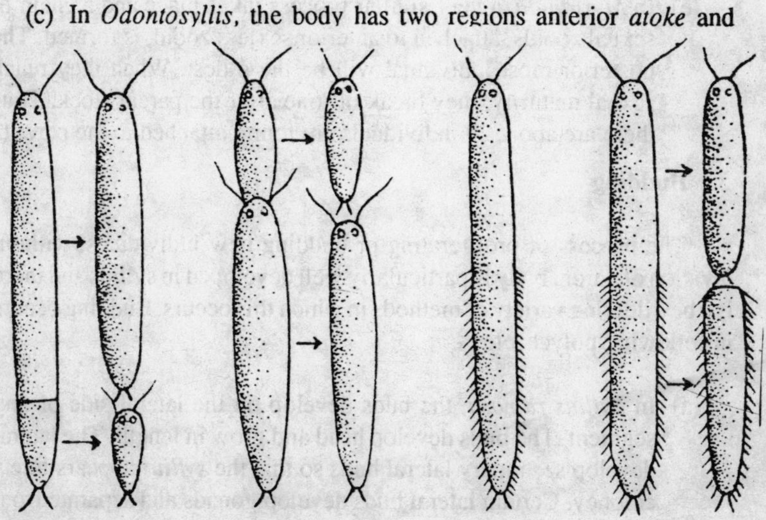

Fig. 11.2 Different types of fission

break off. In *Syllids hyalina* and *Eunice*, the epitoke breaks off but does not develop a head because its independent life is brief, concerned only with swarming. In *Syllids gracilis*, the liberated epitoke may form an incomplete head lacking pharynx and jaws. The anterior atoke portion regenerates the lost part. which again acquires gonads and becomes detached as the sexual epitoke in the next breeding season.

(d) In *Autolytus* the process is more complicated for the formation of a new segments as the new segments begins before the two part of the worm separate so that the posterior sexual worm acquires the head while still attached. In other words a zone of fission is formed and this zone may produced the head of hinder worm but a number of segments which become market out into a series of zooids in which genital organs eventually appear. In this way a chain of sexual zooids attached to the asexual scolex. The zooids are all of the same sex and the posterior of them is oldest and more developed. The male zooids differ from female zooids and formerly related to distinct genous called *Polybostrichus*, the female being known as saconereis and characterized by the possession of ventral brood sac.

(e) In *Myrianida*, a very similar process takes place and a chain of sexual zooids, attached to anterior sexless zooid, is formed. The posterior most individual will be the oldest. When they reach sexual maturity, they break off one after the parent stock. Thus there are about 29 individuals are found attached to the parent.

3. Budding

The process of proliferating or budding new individuals, without division of parent body is particularly well developed in syllids and there is a bewildering variety of methods in which this occurs. Budding occurs in following polychaetes:

(a) In *Syllids ramosa*, the buds develop on the lateral side of any segment. The buds develop head and grow in length. The lateral develop secondary lateral buds so that the *syllius* appears like a coloney. Certain lateral buds develop gonads and separate from the parent.

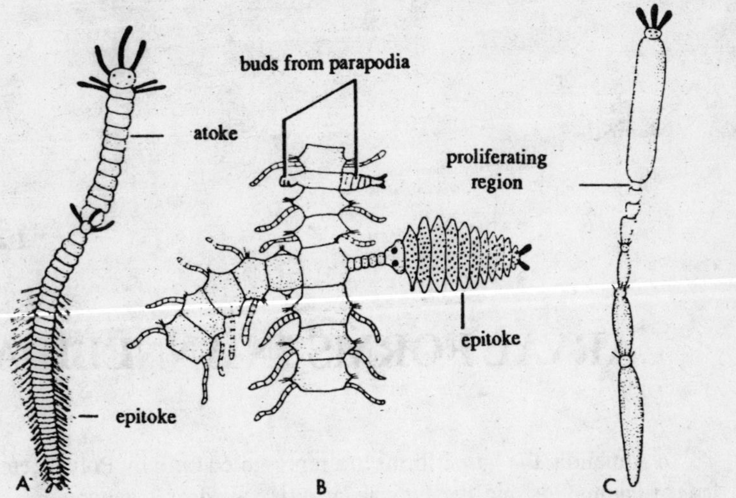

Fig. 11.3 Polychaeta. Asexual reproduction. (A) *Syllis*, with the posterior region forming a reproductive individual. (B) *Syllis*, showing branching of the asexual stock and budding of reproductive individuals from parapodia. (C) *Autolytus*, with a chain of reproductive individuals budded off successively from a proliferating region. (After Borradaile and potts, modified.

(b) *Trypanosyllis* shows great variation from species to species. *I.prolifera* can produce heads from almost any segment and chains of new heads may occur on consecutive segments. *I asterobia* forms clusters of stolons each growing out from a single segment.

Sources of cellular material for asexual reproduction

Asexual reproduction by fission is closely related to regeneration. In both the cases, a part of the body is isolated and reconstructed into a new complete worm. The new structure are formed from formed from certain specialized reserve cells called neoblasts. The neoblasts are actively involved in the development of daughter worms and these are 12actually reported to concentrate in the fission zone.

12

LARVAL FORMS IN ANNELIDA

In Annelida, the larval forms are represented only by Polychaetes and Archiannelids and the typical larva is called as *trochophore*. In Oligochaetes and Hirudinia the larval stage is passed egg resembles the parents in most of the characteristics i.e. development is direct. The trochophore larva is also represented in other phyla such as Mollusca, Phoronida, Echinodermata and Hemichordata.

Structure of typical trochophore larva.

A typical trochophore is free swimming larva. It is ciliated and transparent. It is oval or pear shaped in appearence. Its anterior end is broad and posterior end is narrow. The mouth lies on the ventral side at the anterior end and leads into oesphagus. The oesphagus further bends at right angle to give rise to a bulging stomach. It leads into a long narrow intestine which opens out through the anus. Some of the ectodermal cells become thickened above the mouth region to form a band over which the cilia developed. This band of cilia is called *pre-oral brand of cilia* or *prototroch*. Similarly behind the mouth there is an another band of cilia known as *post-oral band of cilia* or *metatroch*. These cilia beat in a regular manner so as to help the larva in swimming. These cilia give it an appearance of rotating wheel, hence the larva is known as trochophore larva (*Trochos* = wheel; phoros = to bear). In some cases yet another band of cilia may be present encircling the anus and is called as *peri-anal* or *circum anal band of cilia*. At the anterior most end some ectodermal cells get thickened to form an apical plate which bears a tuft of cilia. It also bears two eye-spots and the nervous tissue.

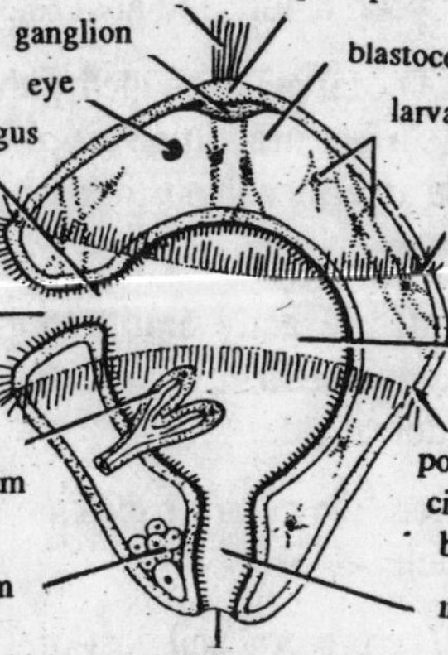

Fig. 12.1 A typical trochophore larva

In addition the larva possesses nephridia.

Larval forms in Polychaeta.

Two types of larva are represented in Polychaeta, one possessing a well-developed blastocoel during the earlier stages of development while the others without a blastocoel. The former is seen in *Pomatoceros* where the ectoderm and endoderm do not touch each other. The an other or later type the ectoderm both lie in contact with each other except at the place where mesodermal band is situated as in *Neries* or *Neanthes.*

Depending upon the distribution of cilia on the surface of body, the larvae are classified as:

1. *Atrochal larva.* Definite bands of cilia are totally absent and cilia are distributed all over the body of larva. Such larvae are rare and represented by *Lumbriconereis*
2. *Prototrochal larva.* Here there is only one band of cilia i.e. pre-oval band, infront of the mouth and this larva is rep resented in the development of *polynoe.*
3. *Telotrochal larva.* Here the larva possesses the pre-oral and post-

oral bands of cilia, but in addition a ring of cilia present around anus. This type of larva is seen in *Spio* and *Nephthys*.

4. *Polytrochal larva*. The larvae are represented with a number of cilia on the body between mouth and anal region.

(i) *Gastrotrochal larva*. The larva is provided with incomplete rings of cilia on the ventral surface in addition to pre-oral and prost-oral bands of cilia situated anterior and posterior to the mouth respectively.

(ii) *Amphitrochal larva*. It is exactly similar to gastrotrochal larva except for the appearance that it is provided with incomplete rings of cilia on the dorsal side.

(iii) *Nototrochal larva*. Here the ciliary bands are present on dorsal side and are incomplete.

5. *Mesotrochal larva*. The pre-oral and post-oral bands of cilia are completely absent, but instead a single or more row of ciliated rings present in the middle of the body of the larva.

Larval forms in Archiannelida.

The larva is a typical trochophore as seen in *Polygordius*. It has broad anterior and narrow posterior end. The apical plate thickening of ectodermal layer at the anterior end and gives rise to the cerebral ganglion and eyes. A strong pre-oral band of cilia is the characteristic feature. A comparatively weaker band of cilia is present on post-oral portion. There is a pair of mesodermal somites present at the hinder end and a nephridium. With the advance of metamorphosis mesoderm gives rise to body segments and the anterior part of larva diminishes in size and modifies to from head - region.

Trochophore Theory. It is based upon the conservation already developed by *Hatschek* (1878). He suggested that the trochophore is a larva of an ancestral from *Trochozoon* which was a common ancestor, at least, a few phyla. According to this theory the trochophore closely resembles a rotifier. However, it is to be noted that the trochophore is regarded as the recapitulating larva of an ancestor but there is no palaeontological evidences for this theory.

Significance of larval form.

Annelidan larvae differ from each other in minor details but the basic plane of their structure is same. Presence of trochophore larva is not the characteristic of phylum Annelida but also occurs in other phyla. The larva is present in the modified form in many branchiopods, bryozoans, molluscs, echinoderms and even in hemichordates. Trochophore larva serves as the source of the stabilizing inter-relationship between different animal groups. Adult annelids and molluscs differ so widely that one can - not expect relationship between these two phyla, but study of embryonic development shows that both have similar stages of their early development and their larval stages.

INDEX